THE
GREAT LAKE
STURGEON

THE
GREAT LAKE
STURGEON

Edited by **Nancy Auer** and **Dave Dempsey**

Michigan State University Press | East Lansing

 Michigan State University Press
East Lansing, Michigan 48823-5245

Printed and bound in the United States of America.

19 18 17 16 15 14 13   1 2 3 4 5 6 7 8 9 10

LIBRARY OF CONGRESS CATALOGING-IN-PUBLICATION DATA
The Great Lake sturgeon / edited by Nancy Auer and Dave Dempsey.
pages cm
Includes bibliographical references.
ISBN 978-1-61186-078-8 (pbk. : alk. paper)—ISBN 978-1-60917-366-1 (ebook)
1. Lake sturgeon. I. Auer, Nancy A. II. Dempsey, Dave, 1957–
QL638.A25G74 2013
338.3'72742—dc23
2012028146

Cover and book design by Charlie Sharp, Sharp Des!gns, Lansing, MI
Cover image of sturgeon eggs on rocks is used courtesy of Nancy Auer and image of
adult sturgeon is used courtesy of Andrew Muir.

g green press Michigan State University Press is a member of the Green Press
INITIATIVE Initiative and is committed to developing and encouraging
ecologically responsible publishing practices. For more information about the
Green Press Initiative and the use of recycled paper in book publishing, please visit
www.greenpressinitiative.org.

Visit Michigan State University Press at *www.msupress.org*

Contents

.

Prefaces

NANCY AUER

It was September 2007. Dave and I were both at an evening reception in the Seaman Mineral Museum at Michigan Technological University. Dave was the guest of honor and had just addressed a public and academic audience on his latest book, which was about the Great Lakes. His passion for the Lakes quickly became obvious.

During the reception, I was introduced as the sturgeon biologist at MTU to Dave and he commented that his next idea for a Great Lakes book was one covering the lake sturgeon. Now my father used to say that I had a rubber face and could hide no emotion, but that night I hoped I had done a reasonable job at hiding my surprise, for my dream was to publish a book on lake sturgeon. I wanted some way to broaden and improve public understanding of and "liking" for this magnificent species, which I had studied for over 20 years. This was a pivotal moment for me, as until then I had found little time to move beyond a dream; writing takes time. I recall standing for a moment looking at Dave and saying to myself: *Speak up, or forever you will lose an opportunity*. So I shared with him my similar desire to publish a book and gain some ground on preserving this great fish. To my surprise he suggested we put our heads together, and that very night he sent me an email with his proposed outline; I still have that email and remain ever grateful Dave said yes to a joint effort.

Our intention with this text is to provide the reader with some history, biology and ecology, and human perspective, as well as suggest some future management ideas. Chapters included in this volume span stories from individuals who are relatively new to appreciating this largest of freshwater fish to those who speak for people whose lives and culture have been intertwined with lake sturgeon for decades or longer. The range of views runs from those who regard the sturgeon as a commercial species to those who wish to carve out a place for the species where it can live out a more natural existence like the wolves and moose of Isle Royale.

Lake sturgeons are endemic to North America, so we have included authors from both the United States and Canada. The chapters are meant to allow readers to discover the inner beauty and mystery of a truly magical fish, one not often encountered or observed, one that some are striving to protect in a sustainable future. Caring for our natural resources takes relationship and respect, which can only be built when we understand how organisms and ecosystems are inextricably linked. We hope you will join us in that journey as you read through these chapters and grow to experience this marvelous fish.

DAVE DEMPSEY

Despite being a native of the Great Lakes Basin and working in the field of Great Lakes policy for a quarter century, I rarely gave thought to the lake sturgeon until the last 5 to 10 years. Fisheries didn't escape my attention—salmon, lake trout, muskie, perch, and other species were impossible to ignore in the realm of policy. These are the sport of the great freshwater fisheries, these are the moneymakers, and these are the source of social and political controversy.

And beginning in 1988, with the discovery of zebra mussels, invasive aquatic species vexed all of us who worked on, or cared about Great Lakes fish. But where was the sturgeon—so large and ancient, yet so invisible to me? My inattention was tantamount to a marine mammal policymaker overlooking whales.

This personal fact is humbling for me to contemplate—and another illustration of the reason the lake sturgeon has had to claw back from near extinction in the last century. Right in front of us but sometimes left out of our policies and our hearts, the sturgeon is what it has always so magnificently been—it is we who are changing, by learning to appreciate it.

That does not mean we will all appreciate it in the same way. Some will study it scientifically, some will incorporate it into their aboriginal lifeways as has been done for centuries, some will simply care for it, while others will want to harvest them sustainably. The sturgeon may be an icon, but each of us will behold a different fish. Each of us will enjoy a different relationship to the sturgeon—and it to us.

In learning much more about the sturgeon in recent years, I have come to admire its resilience. I have also come to admire the people who treat sturgeon as a species worth working to conserve and protect. My coeditor, Nancy Auer, tops the list. She combines impeccable scientific ability, approach, and credentials with a deep regard for sturgeon. This combination of strengths has enabled her to add irreplaceably to the science of the sturgeon and to add incalculably to the Great Lakes community of sturgeon stewards. I am fortunate to know her.

Because of what I have learned about the sturgeon on my own and through the review of the insightful data and feeling expressed about the species in the chapters that comprise this book, I will never again exclude them from my thoughts, my regard, and my sense of their place in the Great Lakes. I hope this book will promote a similar response in its readers. Although the lake sturgeon can rightfully take its place as a biological indicator of the health of the Great Lakes ecosystem, it must also take its place in the passions we all bring to caring for the place we love and protect for future generations.

sections of the country where sturgeon were plentiful, baseballs were commonly made of the eyes of that fish. The eye of a large sturgeon contains a ball nearly as large as a walnut. It is composed of a flexible substance and will rebound if thrown against a hard base. These eyeballs were bound with yarn and afterward covered with leather or cloth. They made a lively ball, but were more like the dead ball of the present than any ball in use at that time" (Morris 2006, 271).

The sum of all destructive forces affecting the sturgeon seemed likely to consign it to memory. "With the drastic decrease in lake sturgeon abundance, most fisheries managers and ecologists believed in the early to mid 1800s that lake sturgeon would eventually disappear as a result of compounding negative pressures. However, lake sturgeon in the Great Lakes proved to be more resilient than previously assumed" (Léonard, Taylor, and Goddard 2004, 232). Indeed.

The survival of the lake sturgeon was synchronous with a remarkable rebound of the Great Lakes. Beginning in the late 1960s, public investments in sewage treatment, enforcement of strict environmental laws affecting business, and dawning environmental awareness among millions of Canadians and Americans reflected in stewardship actions contributed to a remarkable, visible recovery of the Great Lakes.

As algal blooms abated in the lower Great Lakes, introduced salmon replaced unwanted alewives as a dominant species, and bald eagles began reproducing in earnest, the notion of a robust natural world took popular root. "It's amazing how resilient natural processes are once we allow them to work," said my friend, Elizabeth Harris, then executive director of the East Michigan Environmental Action Council in Bloomfield Township, Michigan, in the 1990s. The last decade of the twentieth century offered hope for all life in the Great Lakes after a near ecological collapse at the midpoint of the same century.

The subject of careful census and study since the 1960s also, the lake sturgeon assumed significance before the new millennium began as a bellwether or indicator species for the Great Lakes. Their reintroduction in historic spawning habitats made the sturgeon, as one reporter put it, "a mascot" for a river's recovery (Moule 2008). Evading extinction by clinging successfully to their last viable spawning grounds, the sturgeon, at an estimated 1 percent of its historic numbers, was tougher than the careless despoilers of 100 years before might have thought. In their persistence they appealed to the human heart; and in the light of a new historical narrative they became beautiful.

Sensing the mood, elected officials quickly congratulated their partners and themselves for helping rescue the species. "We are indeed so proud to be part of this international success story of recovery of lake sturgeon in our shared Great Lakes waters," said Canadian member of Parliament Jeff Watson. "It is so heartening to see the amazing success of this sturgeon habitat restoration for the Detroit River International Wildlife Refuge," noted Congressman John D. Dingell. "No one thought this degree of success was possible only 30 years ago" (Friends of the Detroit River 2009).

This is well known. But can a species whose tentative comeback depends not only on expert fisheries biologists but also on the fluctuation in government spending to support their work sustain that recovery? Can the sturgeon of the Great Lakes depend on its human constituency for continuing affection and conservation? The signals are mixed.

The first aquarium in the United States devoted to freshwater fish, the Great Lakes Aquarium in Duluth, Minnesota, opened in 2000 and struggled for years to keep its doors open. Its heavy debt load was one reason, but it also had difficulty attracting a tourist following. Fish of the Great Lakes disappointed some visitors. In 2002, the aquarium's managers "discussed adding saltwater exhibits for more pizzazz, such as bringing in a shark to compare it with a sturgeon" (Associated Press 2002). When Toronto, Ontario, considered its own aquarium in 2005, skeptics said it was a dubious proposal because it would lack "charismatic attractions" like whales, dolphins, and other marine mammals. By 2009, the Duluth facility had an Amazon exhibit and a seahorse display. The Toronto aquarium has not been built.

As much as most anglers, scientists, and nature lovers value the restoration of lake sturgeon for their intrinsic value, poaching is no more eradicated than the species itself. In spring 2009, in plain sight of observers on the Grand River in downtown Grand Rapids, Michigan, a lawbreaker reeled in a five-foot sturgeon and drove away with it in spite of strict regulations (Grand Rapids Press 2009). "Given its size, it would have to be a fairly old fish," Michigan Department of Natural Resources conservation officer Dave Rodgers said. "So losing even one can have an impact on the sturgeon population." The 36- to 40-pound fish was one of only 35 to 40 thought to have spawned in the river in the spring of 2009.

Some of those noting the recovery appear ambivalent. "There are tons of sturgeon in western Lake Erie, especially around the mouth of the Detroit River," said one commercial fisherman in 2005, echoing the complaints of his predecessors 150 years earlier. "They're always tearing the hell out of our nets. I betcha there's a lot more sturgeon there than they can imagine" (Currie 2005).

A counterweight: the enthusiasm of some sturgeon protectors could be contagious, especially with children. The manager of the Detroit River International Wildlife Refuge, John Hartig, called the sturgeon "a show-stopper" for kids. "When you take kids and show them a fish that is five or six feet long, they are blown away. It's a living dinosaur. It's been around that long. They ask, 'How has that thing survived when so many other things have gone extinct?'" (Henry 2009).

The lake sturgeon was also the reasonably charismatic star of a popular IMAX documentary, *Mysteries of the Great Lakes*, which audiences across the region enjoyed in 2008 and 2009. With the haunting steel guitar of Gordon Lightfoot's *Wreck of the Edmund Fitzgerald* in the background and soaring vistas of vast Great Lakes waters as a visual prelude, the movie lovingly portrayed the sturgeon run in a Great Lakes tributary and its human admirers. Elementary school teachers working off the documentary were given materials to help them tell the students "about the water and history of the Great Lakes and some of the aspects that make it an important and unique resource for us all. The common theme throughout all parts of this resource is our Great Lakes friend, Sally Sturgeon. Sally is a lake sturgeon. Sally is over 120 years old. Given everything Sally has been through it is amazing that she has survived so long."

It is; and it is equally amazing that the sturgeon is able to use some of humanity's past mistakes in the Great Lakes as bootstraps for its recovery. A 2004 Michigan Department of Natural Resources report explained a reproducing population of sturgeon in Lake St. Clair's North Channel was assisted by an artificial spawning reef: "The coal cinders at the North Channel site are believed to have been deposited during the late 1800s when coal-burning vessels moored to load salt from a nearby factory and emptied their cinders into the river. The cinder substrate is now zebra mussel encrusted, and the three-dimensional structure of the cinders combined with the zebra mussel layer provide a complex system of interstitial spaces that appears to provide excellent protection for deposited eggs and fry" (Thomas and Haas 2004, 9).

The combination of an unconsciously crafty and opportunistic natural world and human fancy may well perpetuate the sturgeon's good fortune. But while the behavior of nature is largely unalterable, the future of human beings is unknowable. Will our fancy again turn against the fish?

"The outlook for lake sturgeon recovery rangewide is guardedly optimistic," a recent scientific paper observed, "thanks in part to renewed interest in the species, novel approaches to management, new opportunities to eliminate long-standing data gaps, and continued progress in habitat restoration. Recent emphasis on maintaining biodiversity has prompted several new management initiatives to 'bring back the natives . . .' Our guarded optimism, however, is not intended as an 'all clear' regarding threats facing the species" (Peterson, Vecsei, and Jennings 2007, 72–73).

The buffalo/bison may be safe, but the lake sturgeon is not yet. Its populations are tenuous, its reproductive biology a liability, and the affinity of the top beast in the Great Lakes food chain always subject to change. What seems clear is that given half a chance, the lake sturgeon will add to its ancient Great Lakes history.

REFERENCES

Associated Press. 2002. Great Lakes aquarium plagued by money woes. September 9. Http:// www.greatlakesdirectory.org/mn/090902_great_lakes.htm.

Buckley, C. 2009. New love for the endangered uglies? June 29. Http://www.esa.org/ esablog/?tag=charismatic-megafauna.

Currie, P. 2005. Great Lakes legend makes a comeback. Toronto Star, February 22. Http://www. greatlakesdirectory.org/on/022205_great_lakes.htm.

Friends of the Detroit River. 2009. The lake sturgeon have arrived! (news release) May 22. Http:// www.detroitriver.org/fdr-pdf/2009-SturgeonAnncmtFinal_22May09.pdf.

Grand Rapids Press. 2009. State offers $1,000 reward to land Grand River angler who took five-foot sturgeon. June 3. Http://www.mlive.com/outdoors/index.ssf/2009/06/ state_offers_1000_reward_to_la.html.

Henry, T. 2009. Freshwater species making comeback in Great Lakes region. Toledo Blade, October 26. Http://www.toledoblade.com/apps/pbcs.dll/article?AID=/20091026/ NEWS16/910260310.

Kerr, J. W., and F. Kerr. 1864–1892 and 1895–1896. Notes about lake sturgeon in Lakes Ontario, Erie and Huron and associated rivers in the diaries of John W. and his son Fred Kerr, Canadian Federal Fishery officials headquartered in Toronto.

Léonard, N. J., W. W. Taylor, and C. Goddard. 2004. Multijurisdictional management of lake sturgeon in the Great Lakes and St. Lawrence River. In: Sturgeons and paddlefish of North America. G. T. O. LeBreton, F. W. H. Beamish, and R. S. McKinley, eds. Kluwer Academic Publishers.

Michigan Department of Natural Resources. 1973. Michigan Fisheries Centennial Report, 1873–1973. Michigan Department of Natural Resources, Management Report 6.

Michigan State Board of Fish Commissioners. 1888. Ninth report of state fish commissioners.

Morris, P. 2006. A game of inches: The stories behind the innovations that shaped baseball. Ivan R. Dee.

Moule, J. 2008. There's something fishy in the Genesee—again. City Paper, December 24. Http:// www.rochestercitynewspaper.com/news/articles/2008/12/Theres-something-fishy-in-the/.

Peterson, D. L., P. Vecsei, and C. A. Jennings. 2007. Ecology and biology of the lake sturgeon: A synthesis of current knowledge of a threatened North American Acipenseridae. Reviews in Fish Biology and Fisheries 17:59–76.

Roosevelt, T. 1893. The American bison. In: Hunting the grisly and other sketches. Putnam and Sons.

Thomas, M. V., and R. C. Haas. 2004. Abundance, age structure, and spatial distribution of lake sturgeon Acipenser fulvescens in the St. Clair system. December. Http://www.dnr.state. mi.us/publications/pdfs/ifr/ifrlibra/research/reports/2076rr.pdf.

U.S. Fish and Wildlife Service. 2009. Great Lakes Lake Sturgeon Web Site. Http://www.fws.gov/ midwest/sturgeon/biology.htm.

Wildlife Conservation Society. 2008. Survey says: Let bison roam. November 18. Http://www.wcs. org/new-and-noteworthy/survey-says-let-bison-roam.aspx.

Wright, R. R. 1892. Preliminary report on the fish and fisheries of Ontario. In: Commissioners' report, Ontario Fish and Game Commission. Warwick & Sons. (Henry A. Regier deserves thanks for unearthing this report.)

NANCY AUER

Form and Function in Lake Sturgeon

THE LAKE STURGEON, *ACIPENSER FULVESCENS*, WAS FIRST DESCRIBED IN 1818 by a botanist from Turkey named Constantine S. Rafinesque. He encountered lake sturgeon during a survey of the flora and fauna of the Ohio River (Rafinesque 1820). The lake sturgeon is the only endemic sturgeon of the genus *Acipenser* found through-out the three closely related, freshwater drainage basins in North America, those of the Mississippi River, Great Lakes, and Hudson Bay (Ferguson and Duckworth 1997). Eight other species or subspecies of sturgeons in two genera are recognized in North America (table 1).

The sturgeons are one of the oldest fishes on earth, bridging evolutionary time between the closely related and similar-looking sharks (fishes with full cartilaginous skeletons) and the early true bony fishes first represented by the gars and bowfins (figure 1). Sturgeons possess some bone in the form of plates, called scutes, on their body surface and head that have persisted in the geologic record (plate 1). Fossils of sturgeon scutes date to between 100 to 200 million years of age, placing these fish on earth during the age of dinosaurs, the late Cretaceous (Hilton and Grande 2006).

Table 1. North American Sturgeon Species Ordered by Known Maximum Size

LOCATION IN NORTH AMERICA	COMMON NAME	SCIENTIFIC NAME	MAX. AGE (YEARS)	MAX. TOTAL LENGTH (MM/IN.)
West Coast	White	*Acipenser transmontanus*	> 80	6500 / 2560
East Coast	Atlantic	*Acipenser oxyrinchus*	60	4100 / 161
Midwest	Lake	*Acipenser fulvescens*	154	2800 / 110
Gulf Coast	Gulf	*Acipenser oxyrinchus Desotoi*	30*	2400 / 94*
West Coast	Green	*Acipenser medirostris*	> 45[†]	2000[‡] / 79
Midwest Rivers	Pallid	*Scaphirhynchus albus*	40	1900 / 75
East Coast	Shortnose	*Acipenser brevirostrum*	67	1200 / 47
Midwest Rivers	Shovelnose	*Scaphirhynchus platorynchus*	27	1000 / 39
Alabama & Mississippi	Alabama	*Scaphirhynchus Suttkusi*	unknown	762 / 30[§]

Sources: Data from LeBreton, Beamish, and McKinley 2004 unless otherwise indicated.
*http://www.flmnh.ufl.edu/fish/Gallery/Descript/gulfsturgeon/gulfsturgeon.html
[†] http://www.krisweb.com/biblio/klamath_usfws_nakamotoetal_1995_sturgeon.pdf
[‡] http://www.nmfs.noaa.gov/pr/pdfs/species/greensturgeon_detailed.pdf
[§] http://library.fws.gov/Pubs4/alabama_sturgeon.pdf

Figure 1. Large adult lake sturgeon at spawning time, Houghton, Michigan 1990. [Photo by N. Auer.]

Why Have Sturgeon Survived for So Long?

The 27 species of sturgeons known to live in the Northern Hemisphere have survived due to a combination of several life history strategies. Most sturgeons fill an ecological niche by retaining, from their shark ancestors, several features that have persisted over time, withstood the power of natural selection and proven beneficial to survival (Auer 2004). The sturgeons have combined several life history strategies, which include the following:

- Large body size and shape, large muscle segments and oil-rich connective tissue that assist buoyancy and energy-efficient swimming
- Early life development of fast growth and sharp boney scutes that help deter predators
- Maturation of reproductive organs later in life and intermittent spawning
- Feeding on benthic organisms and organic material, which allow a more passive feeding strategy, requiring less physical energy than a typical predator
- Occupying the light-limited, bottom waters of lakes and large rivers that protect them from predators and unusual fluctuations in water temperature
- Living to a great age, which allows them, over short or long time periods, to persist through flood, drought, warming, or cooling events.

Fishing pressure, barriers, and dams blocking spawning migration routes and persistent chemical contamination are human impacts that have and continue to reduce the future success of this unusual organism in its remaining habitats. Let's take a closer look at some of these life strategies.

Body Plan

Sturgeons represent the transition between fishes with only cartilage, like sharks and rays often called elasmobranchs, and the true bony fishes most often eaten or caught in sport fisheries today. Sturgeon keep a similar body design to that of the shark, but instead of retaining oil in a large liver, as do the sharks for buoyancy, these fish possess large muscle segments and a rich oil between these segments that was either boiled off or smoked off prior to consuming the flesh (Harkness and Dymond 1961). The sturgeons also possess an air bladder that produces a gelatinous substance called isinglass that was used in clarifying beer and wine and in jam and jelly preparation (Harkness and Dymond 1961). The oil, muscle, and air bladder provide the body

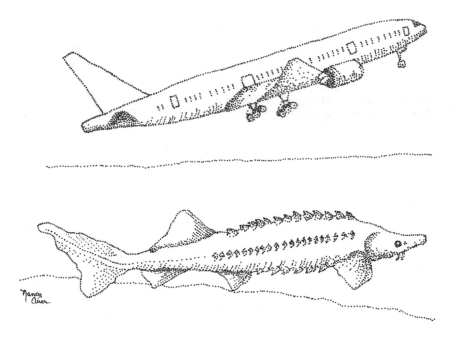

Figure 2. A sturgeon lifting off the bottom of a lake or stream might resemble an airplane at takeoff, using downward pressure by the tail and lift generated by pectoral fins as it moves forward. [Drawing by N. Auer.]

structure and buoyancy needed to move the fish without a bony internal support system. The bony plates they do have on the head and along the sides of the body in five rows are believed to be for protection, as the scutes have razor-sharp edges when the fish are very young. The scutes become dull and rounded with age.

Sturgeons spend most of their lives cruising just above the lake, ocean, or river bottom, and must journey upstream during spring to spawn. However, they have adapted to using the currents and their body composition (high oil content and large air bladder) to their advantage for movement. A sturgeon starting to swim from a stopped position looks just like an airplane at the time of takeoff (Wilga and Lauder 1999) (figure 2). The sturgeon tail is broader on the top than at the bottom (termed heterocercal), and a single tail beat will actually push the back end of the fish down. Combine this with a small push up from the unusually large pectoral fins, and the streamlined head lifts into the current just as a plane takes off into the wind. Small energy-efficient pushes allow the sturgeon to use very little energy to lift itself into movement. The shark and sturgeon body plans are some of the most efficient for cruising over bottom substrate.

Modern fishes such as salmon and sunfish can regulate their depth position in the water by direct secretion of gases into the air bladder from the blood. Since sturgeons are more primitive, they lack this ability and must gulp air to keep their air bladder inflated. When they inhabit deep water for any period of time they begin to lose air from the bladder due to hydrostatic pressure (weight of the water column above the fish creates pressure as scuba divers experience) so they rise or even rush to the surface and can be seen rolling or jumping to obtain air (Watanabe et al. 2008).

The sturgeons have also kept impressive sensory development to achieve their successful life history strategy. Since these fishes live in deep, dark, turbid waters, they are believed to be nearsighted, yet they are sensitive to light. When very young, most sturgeon are photonegative (move away from light), often seeking out dark crevices and holes during daylight hours. Even as they mature they seem to avoid strong light, which is advantageous as potential predators less easily see them.

Some features of the sturgeon eye confirm its place in the evolutionary tree between sharks and bony fishes. Sturgeons possess a feature typical of many early fish such as sharks called the tapetum lucidum (Nicol 1969). This name refers to the special cells, filled with a crystalline material in a layer just behind the retina in the eye, that reflect light back into the eye and help concentrate light and improve vision at night or in murky waters. This layer is common in many animals but obvious in sharks, cats, and deer.

Sturgeons also have a simple retina dominated by rods and with single cones, in contrast to what is seen in later bony fishes, which have a more complex retina with multiple cone photoreceptors (Rodriguez and Gisbert 2001, 2002; Sillman et al. 2005). Eyes with mostly rods are generally found in animals adapted for life in low-light conditions (Sillman et al. 2005). The arrangement of the pigment in the sturgeon cornea also helps reduce eyeshine upward out of the pupil, making the eye less conspicuous when viewed from overhead (Nicol 1969), where other predatory fishes may be cruising.

Sturgeons also have well-developed smell and taste receptors located in their nostrils and on the surface of the barbels and bottom of the head near the mouths (plate 2b). As they cruise along the lake and river bottom, they use their barbels to gather sensory information on what is on or near the substrate surface (Harkness and Dymond 1961). Sturgeons are known to consume all manner of organic debris, typically eating worms, small clams, crayfish, and decomposing fishes on the lake bottom. Some studies have even found sturgeon stomachs to contain items such as onions, corn or wheat from grain elevator or railroad spills, and cigarettes (Harkness and Dymond 1961). Recent reports indicate that adult sturgeons may be one of the few organisms that consume the exotic invasive zebra mussel (U.S. Geological Survey 2008).

Sturgeon are known to be curious, often coming to investigate ice-fishing decoys (Harkness and Dymond 1961). Whether they see or sense these objects through sound or vibration traveling through the water and sensed through their large air bladders has yet to be determined. Sturgeon are also known to be sensitive to electrical transmission, migration routes possibly influenced by overhead transmission lines (J. Hayes, SUNY ESF, pers. comm).

Growth

Sturgeon are typically thought to be slow-growing fishes, but during their first year of life grow quickly, increasing in length from about 25 mm (one inch) at hatch to 125 mm (five inches) at about five months of life. Young hatch from small, ⅛-inch-diameter eggs spawned in the springtime in the rapids of large river systems (plate 3). The eggs and sperm are released together by the adult male and female fish, and fertilized eggs roll and adhere to the undersides of clean rocks. Depending on water temperature, the eggs will hatch in about five to eight days. Once hatched, the little fish burrow into the river gravel where it is dark and for a few days slowly absorb the yolk in the yolk sac remaining from the egg (plate 4). Once the yolk is gone, the little fish must begin to find food and start moving from the spawning location by rising out of the gravel at dusk and drifting with the current for a few miles before settling again and feeding and hiding during the daytime. This drift can occur for great distances until the young find a rich organic region within the river or near the river mouth in which to feed. Some young stay in rivers to feed, while others appear to move out into connecting lake systems, usually systems with large productive deltas. Some characteristics of sturgeon early in life, such as what they feed upon and exactly where they spend time, are still being investigated.

Lake sturgeons will utilize both river and lake environments once they grow beyond their first year of life. They are usually found over sand or sand and pea gravel substrates and seem to feed on organic matter and organisms drifting with the currents (Harkness 1923). They remain juveniles, much like humans, for almost 15 to 20 years before they reach sexual maturity and return to natal rivers to spawn.

Reproduction

All sturgeon species are believed to spawn in their natal river, locating it by imprinting to chemical cues during early life much like salmon and trout. It is amazing that a fish 15–20 years old can return to a river after such a long absence. This is one of the reasons it is important to keep river corridors unblocked and river environments clean and healthy for fishes that must find their way back to natal spawning areas.

In northern Michigan, lake sturgeon males reach maturity at about age 15 or 130 cm TL, while females take a bit longer and might be 20 years old before they first spawn at about 140–150 cm TL (Auer 1999). This strategy is beneficial as it increases the probability that male and female siblings will likely not spawn together in the same years and thus limit genetic variability. Males are found to spawn about every other year, while females spawn once every five to nine years. That is because it takes time and energy for a fish to produce eggs. A 25-year-old, 50-pound female may contain 250,000 eggs (Harkness and Dymond 1961), the resources for which are slowly accumulated over the long period away from the spawning river. The loss of one female through accident or poaching, therefore, can drastically impact a small, remnant stock of sturgeon.

Producing and spawning a single large batch of eggs over a long and intermittent time frame is beneficial as a reproductive strategy for a fish under natural circumstances. Migrating to and spawning in rivers for fishes exposes them to possible predation, usually means food is limited, and costs a great deal of energy. By spawning in a stream in which they were originally spawned, the need to expend energy searching for adequate habitat is reduced. The site where they were spawned most probably had adequate water flow, temperature, and oxygen for egg development. The adults become concentrated in the rapids areas to spawn, so mate finding is easy. They depart the river area quickly after spawning, which reduces any possible predation on eggs of their own species as well. However, this strategy of reproduction can be compromised when rivers are influenced by human and industrial discharges, water withdrawals, or barriers.

Feeding

Sturgeon feed along lake and river substrate, detecting organic matter with their four barbels, which are positioned in front of the mouth (plates 2a and 2b). They have no teeth, so they actually suck organic material and organisms, along with muck and sand, into the mouth. The sand and muck is rejected out the gill openings, while

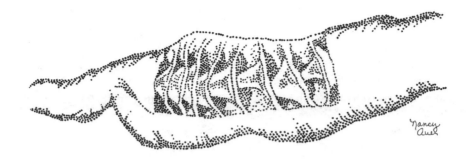

Figure 3. View of spiral valve in sturgeon stomach, showing continuous movement of material through the stomach, increasing absorption area. In this rendering, the fish mouth would be at the top of the image, and the narrower passage at the bottom would be close to the anus. [Drawing by. N. Auer.]

the organic material is swallowed. Sturgeons have a unique stomach with what is known as a *spiral valve* (figure 3), which reduces the need for a long gut to slowly digest food, thus saving internal space for energy storage and reproductive organs. Food in the spiral valve gut circulates over a spiral path, similar to that of a spiral staircase; the gut itself is full of folds to increase surface area and absorption of nutrients. Sturgeons are passive, opportunistic feeders. They usually don't search out and track down organisms, but rather situate themselves in rich environments (lake shorelines, river mouths, wetlands, etc.) and feed on what is encountered there. This strategy requires less physical energy than that of a more aggressive predator like a salmon or trout seeking forage fishes on which to feed, so energy can be conserved and placed into growth and reproductive products.

Feeding on a diverse diet of various organic products is also a good strategy for survival. When we think of food webs or food chains we often think of the typical piscivore (fish eating) or herbivore (vegetation eating) predator/prey dependent cycle, such as that of the lake herring or the cisco feeding on zooplankton or lake trout feeding on rainbow smelt. However, if for some reason such as disease or overharvest, the forage item (small fish or vegetation) become less abundant, then the specialized predator will also decline in number. A diverse diet, low on the food web, allows the lake sturgeon to find some type of food throughout almost the entire year.

Habitat

Most sturgeons select for environments with less light intensity than that found in surface or clear, open water. These fish are often found in large turbid or tannin-dark lake and river systems. The lake sturgeon is found throughout the Great Lakes, Mississippi, and Hudson Bay watersheds, yet is rarely encountered by people simply because to save energy sturgeon cruise along the lake and river bottoms.

Age

Lake sturgeon are known to reach great size and age. Two "record" sturgeon have been reported from the Great Lakes, one from Lake Superior in 1922 weighing 310 pounds and over seven feet long (Scott and Crossman 1973) and one from Lake Michigan in 1943 of about the same size (Becker 1983). These large sturgeon are believed to be over 150 years old (Scott and Crossman 1973).

A life history strategy often found in large-bodied organisms is the ability to live to great age. Sturgeon are similar to humans and some mammals in that they reach reproductive readiness over a time frame of many years. This can be a life history advantage over smaller-sized creatures for many reasons, and may explain why they have not changed much physically over the millennia. First, they can persist over time by remaining in deep cool water or large bodies of water, where their metabolism can slow and energy demands become low. This could occur during chaotic global events such as cooling, drought, flooding, and heating, which can result from climate change. If for some reason they cannot spawn because of conditions that are unfavorable for reaching the spawning site, they can live to return in future years to try again.

Overview

Sturgeons are unique species of fish, and in North America the lake sturgeon is the largest, most widely distributed freshwater fish in inland freshwaters. This species is a historic fish; it represents one of the steps in evolution between fully cartilaginous fish (i.e., sharks and rays) and the more familiar ray-finned or "bony" fish (i.e., bass and salmon). Sturgeons have retained the form and sensory adaptations of the sharks yet show the beginnings of bony fish by the presence of protective scutes. These special fish can live to a great age, can grow to a great size, and have persisted for thousands of years. In fact sturgeons join the ranks of such aquatic creatures as horseshoe crabs and sea turtles, thriving pretty much unchanged in form and function since the days when dinosaurs ruled the world. Human actions now put sturgeons at risk of survival by a combination of overfishing, barricading of corridors needed to reach spawning sites, and destruction of important spawning, feeding, and resting habitats.

REFERENCES

Auer, N. A. 1999. Population characteristics and movements of lake sturgeon in the Sturgeon River and Lake Superior. Journal of Great Lakes Research 25(2): 282–293.

———. 2004. Conservation. *In*: Sturgeons and paddlefish of North America. Greg T. O. LeBreton, F. William H. Beamish, and R. Scott McKinley, eds. Kluwer Academic Publishers.

Becker, G. C. 1983. Fishes of Wisconsin. University of Wisconsin Press.

Ferguson, M. M., and G. A. Duckworth. 1997. The status and distribution of lake sturgeon, *Acipenser fulvescens*, in the Canadian provinces of Manitoba, Ontario, and Quebec: A genetic perspective. Environmental Biology of Fishes 48:299–309.

Harkness, W. J. K. 1923. The rate of growth and the food of the lake sturgeon (*Acipenser fulvescens* LeSueur). Univeristy of Toronto Biological Studies Series 24. Publication of the Ontario Fisheries Research Lab 18.

Harkness, W. J. K., and J. R. Dymond. 1961. The lake sturgeon. Ontario Department of Lands and Forests.

Hilton, E. J., and L. Grande. 2006. Review of the fossil record of sturgeons, family Acipenseridae (actinopterygii: acipenseriformes), from North America. Journal of Paleontology 80:672–683.

LeBreton, G. T. O., F. W. H. Beamish, and R. S. McKinley, eds. 2004. Sturgeons and paddlefish of North America. Kluwer Academic Publishers.

Nicol, J. A. C. 1969. The tapetum lucidum of the sturgeon. Contributions in Marine Science 14:5–18.

Rafinesque, C. S. 1820. Icthyologia ohiensis. Author.

Rodriguez, A., and E. Gisbert. 2001. Morphogenesis of the eye of Siberian sturgeon. Journal of Fish Biology 59:1427–1429.

———. 2002. Eye development and role of vision during Siberian sturgeon early ontogeny. Journal of Applied Ichthyology 18:280–285.

Scott, W. B., and E. J. Crossman, 1973. Freshwater fishes of Canada. Bulletin 84. Fisheries Research Board of Canada.

Sillman, A. J., A. K. Beach, D. D. Dahlin, and E. R. Loew. 2005. Photoreceptors and visual pigments in the retina of the fully anadromous green sturgeon (*Acipenser medirostrus*) and the potamodromous pallid sturgeon (*Scaphirhynchus albus*). Journal of Comparative Physiology A: Neuroethology, Sensory, Neural, and Behavioral Physiology 191:799–811.

U.S. Geological Survey, Great Lakes Science Center. Zebra mussel. 2008. Http://www.glsc. usgs.gov/main.php?content=research_invasive_zebramussel&title=Invasive%20

Invertebrateso&menu=research_invasive_invertebrates.

Watanabe, Y., Q. Wei, D. Yang, X. Chen, H. Du, J. Yang, K. Sato, Y. Naito, and N. Miyazaki. 2008. Swimming behavior in relation to buoyancy in an open swimbladder fish, the Chinese sturgeon. Journal of Zoology 275:381–390.

Wilga, C. D., and G. V. Lauder. 1999. Locomotion in sturgeon: Function of the pectoral fins. Journal of Experimental Biology 202:2413–2432.

JIMMIE MITCHELL

N'me

I HAVE BEEN ASKED IN A GOOD WAY TO SHARE THE INFLUENCES AND PERSPEC-tives driving modern tribal resource management. In honoring this request, I am proud to share with you a project that is a premiere expression of modern-day tribal sovereignty, the *n'me* restoration project of the Little River Band of Ottawa Indians.

Before I attempt to describe this project in greater detail, let me first explain the inspiration utilized in the creation of our sturgeon management plan, an Indigenous belief structure known as *Baamaadziwin*, which translates into "living in a good and respectful way."

The Anishinaabek (the name in our language we refer to ourselves as) have a belief system that has been in existence since time immemorial and has been passed on in the oral tradition from generation to generation to our present times. When we seek a teacher in Baamaadziwin, it is our understanding that we will receive more than basic instructions in a "good and just way to live." As these teachings are learned, we are also introduced to the guiding spirits of Baamaadziwin, spiritual guides who will assist us on our new path till the end of our days on earth.

This connection between spirit-world and our own is not obtainable by research-ing books or visiting the World Wide Web. The introduction is made in the presence of teacher with student, connecting them to the unseen powers surrounding us all, a procession of knowledge and faith that has occurred since the beginning of our existence. I will explain later how this influence transcends the Anishinaabek understanding of the world, and how so much of what we understand has been learned from the wisdom of the wilderness.

The spirit that is connected to our belief system guides the Anishinaabek to our respective responsibilities, beyond that of being "good and just" people—to being servants, devoting ourselves to making a difference in all that has occurred and may still be occurring within our respective communities and environment. This community-minded service for some of us also includes restoring the balance of our shared natural environment and of all inhabitants who are dependent upon a robust ecosystem.

As is told in the creation story of the Anishinaabek, Giizhemanido (The Creator) put into existence all of the life-forms we have come to know. Some 360 million years ago, during the creation of the first n'me, Giizhemanido also decided to create the very first Anishinaabek from the remains of the first animals that had lived and died.

In Giizhemanido's infinite wisdom, the positive attributes that he admired most in those first animals thusly became inherent in the Anishinaabek; and therefore our *dodem*, or clan, system was created and set into motion.

The phrase *n'dodem* in the language of the Anishinaabek literally translates into English as "I have the heart/spirit of . . ." A sturgeon clan person, for instance, would declare his or her clan in this way: *n'me n'dodem*, or "I have the heart/spirit of a sturgeon."

This introduction is very significant in the cultural interactions of the Anishinaabek because as we venture along this journey, living connected to the harmonious ways of Baamaadziwin, whenever we connect with other Anishinaabek, it is respectful to explain who we are, where we come from, and, if known, which clan we represent.

Following the cultural practices of the Anishinaabek, clan responsibilities are expressed in fundamental rules to abide by. For instance, clan people are instructed not to marry into their own clan. This was done as a means of maintaining genetic diversity throughout the various Anishinaabek communities. Clan people are also charged with various tasks and responsibilities within the communities; some possess the role of teachers, others protectors, hunters, and healers, and through the clan system a well-balanced community was ensured.

Upon the arrival of the dominant society, its effects began to negatively tilt the delicate balance of our world, including that of the clan system. In the throes of genocide, clan leaders were targeted as a threat opposing the emerging strides across the landscape driving a new concept known as Manifest Destiny.

In the wake of the losses occurring to the world we knew, loved, and depended upon, tribal leaders opposing these changes were dealt with severely, and most often with extreme prejudice.

As the balance of the environment teetered out of sync, new concepts like "resource management" and "public trust doctrine" were coined to leverage the needs of the dominant society against those of the Anishinaabek, causing negative environmental consequences that reverberate into our current landscape. Likewise, with the loss of leadership, land, and naturally occurring sustenance, the once vibrant cultural kinship between earth, animal, and the Anishinaabek wilted.

To explain this detriment in greater detail: our healing ways suffered, as our doctors were no longer assisted by their clan animal counterparts who used to confer their healing powers on them, a significant component to our faith. The naturally occurring medicines we had relied upon to heal illnesses since the beginning of time no longer grew on the degraded soil, parched by the absence of the forest canopy, torn off by clear-cutting and subsequent burning. New diseases inflicted upon the Anishinaabek killed scores of our people, forcing us to turn toward the medicine of the dominant society.

The loss to the indigenous animal and human populations within the Great Lakes Basin since first contact is profound. The dimension of suffering we have known and continue to know is reflected in the degraded condition of our environment, yet we still consider her sacred, so much so that we refer to her as *Wegemind Aki*, or Mother Earth.

As our tribal nations begin to grow strong again, we haven't forgotten our primary role: To mend the circle of life that has fallen so far out of balance, we must rely upon our codependency with our clan species in conjunction with the roles that were bestowed upon us by Giizhemanido.

During the treaty negotiations of 1836, our chiefs consciously argued for language that secured the usual privileges of occupancy the Anishinaabek relied upon, then as now. The treaty was negotiated to pave the way for our lands to become a state; the negotiations were brought forward to divert a war about to break out between the United States and the Odawa, Ojibwe, and Bodwewadomi nations, the Three Fires Confederacy. In their wisdom, our chiefs knew that if they could continue to occupy our homelands and follow the natural order that Giizhemanido had granted us in this paradise that we called home, we could continue to exercise our unique identity as Anishinaabek.

Then, shortly after the treaties were signed, the unthinkable occurred: the army took our children away from us by force and sent them to Indian boarding schools across the country, where they were taught in the cruelest of ways to be ashamed of themselves and our society as backward and lacking intellectual capacity. We had been stripped of our connection to the lands, animals, and medicines. For the next 160 years, the Anishinaabek languished as our families clung to the slim faith that one day we would return to our rightful place as a proud and unique society. Despite the intentional disenfranchisement from everything the Anishinaabek knew, loved, and held sacred, we endured.

Remaining from the tenets of a nation that once stood as the proud original caretakers of the lands, the beliefs we clung onto eventually fanned the embers of our once great nation into flame, a flame that has illuminated our path, calling us home to begin anew in our rightful place in Creation.

In 2003, the treaty of 1836 was called into question in the Sixth District Federal Court, targeting specifically the language that preserved our usual privileges of occupancy until the land was required for settlement. The usual privileges of occupancy during the 1836 treaty included hunting, fishing, and gathering practices, but certainly more with the economic change taking place within the region. The Tribe had already before 2003 been relying on the language preserved by our chiefs in 1836 to hunt, fish and gather, but also in our efforts to restore sturgeon populations in the Big Manistee River and to initiate economic development to fund the endeavors of the Nation.

After an exhausting and costly negotiation process that lasted nearly four years, in the fall of 2007, the Tribes were able to reaffirm the original language in the1836 treaty, securing the rights not only to hunt, fish, and gather, but also to conduct co-management initiatives, including the restoration, reclamation, and enhancement of species and habitat that are of cultural significance to the Tribe.

These inherent rights serve not only the Tribe; their benefits are to be shared with people from all races of life. In keeping with our Seventh Generation teaching, we have to ensure that the next seven generations of the people yet to be born inherit the same benefits that we enjoy and utilize today.

Despite the advancements in resource management, the continued progress of humans continues to create a detriment to all forms of naturally occurring life. For example, it is difficult for me, as an Indian person, to drive my vehicle and see the massive number of animals killed along the road, animals that do not possess the wherewithal to comprehend the physics involved in cars speeding toward them.

In the teachings that I have received, I've been taught to ask for forgiveness for doing something wrong but also to take steps to make amends. The Tribe, through its sturgeon-rearing program, is striving to make amends for 170 years of destruction done against this noble species.

To qualify this detriment in regards to sturgeon populations, we have to travel back to the time when the last of the great forests were cut in Michigan. After the last of the trees were felled, the steamships had no fuel left to burn in order to propel these vessels across the lakes. During the sturgeon runs of the time, nets were stretched across the rivers. The caught sturgeon were dried whole and stacked like cordwood and subsequently burned as a new source of fuel, at least until the population was decimated, in the same fashion as the great forest.

A current harm to the remaining sturgeon population comes with the massive number of lake trout and salmon fry that are stocked in the rivers in the spring of each year, about the same time the sturgeon larvae begin their drift. At this time, tribal biologists enter the dark waters of the river around midnight and painstakingly collect the elusive larvae from the frigid waters of the Big Manistee. The larvae are then carefully transported to the Tribe's patented streamside rearing facility, where they are fed, monitored, treated for disease, and raised until their protective plates are formed.

The process takes approximately four months, when tribal biologists feel the sturgeon survival rate is ensured. The Tribe's annual sturgeon release ceremony takes place each September along the shores of the same river the larvae were retrieved from, the source that we feel imprints them with the water of the Manistee.

Miigwetch (thank you) for listening to our story of n'me, and our humble attempts to preserve this amazing being that is loved and revered by so many of you.

HENRY A. REGIER, ROBERT M. HUGHES, AND JOHN E. GANNON

The Lake Sturgeon as Survivor and Integrative Indicator of Changes in Stressed Aquatic Systems in the Laurentian Basin

Our Laurentian Basin is ancient geologically, and the lake sturgeon lineage in our basin is ancient biologically. During the last five centuries, much of the southerly half of this basin's waters has been transformed—mostly by humans of European origin—from a vast, clean, cascading "riverine system" to clotted strings of confined and dirty reservoirs and lakes with deformed tributaries and "connecting channels," that is, into what we may term a "reservoirine system" (see box 1).

The living part of the ecosystem in most of the southerly waters has changed from intricately organized and elaborately choreographed systems of native, freshwater, mostly riverine taxa to a highly altered and relatively disordered state with unwanted alien taxa from shallow seas and brackish bays far away. But that long-lasting trend may not be our destiny. As complex self-organizing systems, these waters and their sturgeon populations retain resilient propensities in spite of their old age and man-caused disabilities.

There have been three kinds of early waves of entry of unwelcome alien creatures into our basin:

Box 1. Reservoirine Alterations

In 1971 some of us of an older generation convened a Symposium on Salmonid Communities in Oligotrophic Lakes (Loftus and Regier 1972). We compared and contrasted the case histories of numerous lakes that were once dominated by salmonine associations to see whether we could infer some major effects of three cultural stresses: poor fishing practices, environmental pollution, and introduction of undesirable alien species. The lakes that we selected fell into three sets; lakes at the margin of the Laurentian Shield in North America including the Great Lakes; deep Fennoscandian lakes; and lakes of the European Alps. During our comparative consideration within and between these three sets of lakes we noted that researchers on the Alpine lakes added an additional important stress related to hydrographic and hydrological modifications that they called *Verbauung* in German. We found that this term, which often had a pejorative connotation, meant something like "human alterations that obstructed or debased natural physical processes purportedly for the immediate benefit of the human obstructors/debasers." What we refer to here as "reservoirine alterations" as a set has features like those that follow such *Verbauung*.

A key feature of "reservoirine alterations" is our culture's primary focus on the physical mass of water in aquatic ecosystems. (The aquatic ecosystem reaches into soils and aquifers and perhaps into local weather patterns, etc.) Nested within this primary focus on the mass of water are secondary physical foci: water level and its fluctuations; water flow and its fluctuations; copious fluid in a depression in a landscape for dumping wastes some of which were carried downstream and out of sight by currents; concentration of pollutants such as acids that corrode engineering works; temperature of the water; and so on. In recent decades Sproule-Jones (e.g., 2002) has described the role of historic preemption by use and by law of features of our natural ecosystems. Such preemptive practices that permeate current realities in our Laurentian Basin can be traced back to ancient history in Europe and Asia. With each major

- The Europeans themselves with their effective invasion strategy that included explorers, soldiers, missionaries complemented by traders, and settlers strongly influenced by an exploitative-consumptive culture
- The human diseases that Europeans unintentionally brought with them
- The alien terrestrial and aquatic species, including domesticated livestock and

technological initiative like this, prior residents of the affected lands, including Aboriginal peoples, likely were harmed deeply in ways for which they received poor compensation at best.

Preemption of water as physical mass relates to old-fashioned "progressive" industrial interests with respect to our aquatic ecosystem. Here "industrial" includes water for agriculture, including irrigation and drainage; cleansing and cooling things; electrical services including potential energy and cooling liquid; and floating and passing commercial vessels. At a macro level, the U.S. Army Corps of Engineers, the U.S. Bureau of Reclamation, and Public Works Canada serve such interests and the preoccupations of these federal institutions dominate and suppress the interests that are not primarily and secondarily as sketched above. The "obstructive" practical engineering methods that follow from such a set of primary and secondary foci have led to a transformation of a pristine riverine ecosystem to a debased reservoirine ecosystem that has adapted to old and crude industrial practices.

The leading initiatives of the older basinwide governance institutions (the binational International Joint Commission and Great Lakes Fishery Commission, the Great Lakes Commission with its connections to Canadian provinces) all explicitly or implicitly accept the preeminence of what we have termed primary and secondary focal interests above. So "reservoirization" is a meta-adaptation of our basin's aquatic ecosystem to old-fashioned obstructive industrialism. The reservoirization syndrome includes damping floods and droughts of natural flow regimes that reset the dynamics of water bodies but are troublesome to human enterprises; destroying wetlands, riparian zones, complex channels, large woody debris, floodplains, and temporary streams and ponds—all key to natural ecosystem processes; disconnecting surface and ground water, leading to salt deposition in xeric areas and flash flooding in others; and severing critical migratory routes for aquatic species such as fish and mussels.

plants, which were deliberately or accidentally introduced as a result of various kinds of stocking and barrier removal

Humans, mostly of European descent, used and abused our basin's waters in countless ways. The historical sequence of initiation of various human activities with major adverse ecological effects took place approximately as follows: the trapping of beaver and removal of the impermanent dams, ponds and the wetlands that they created; physical restructuring of streams and shorelines through snagging, more-permanent damming, channeling, and hardening of banks; fishing exploitatively; loading of organic wastes and sewage; introduction of alien fish deliberately or accidentally; loading of plant nutrients; loading of toxic chemicals, then radionuclide physicals, then trace organo-contaminants and pharmaceuticals; introduction of a broad spectrum of alien species mostly via ships' ballast waters; acid rain; and change in the climate. Each of these classes of stress, as well as any transformational synergism among stresses, has likely affected adversely the numerous lake sturgeon populations in these waters. Nevertheless some sturgeon populations did survive, in small numbers, and are now showing signs of recovering resiliently.

We, the three authors, are of European descent ourselves. As "Europeans" we wish to lay no claim to being the worst ever of the world's conquerors throughout history, noting the work by Diamond (1997) on this subject. But here in the Laurentian Basin, half a millennium ago, it was Europeans who entered as external invaders and eventually affected ecological and other aspects of the basin's history deeply, if often in ways that would have been unintended had they given the matter some ethically informed thought. With our broad-brush approach in this heuristic chapter, we will use the term "European" without making an effort to strike a fair balance, from some distant and disinterested perspective, between goods and bads that Europeans brought with them.

All of the phenomena mentioned above—except sea lamprey as an alien invader, acid rain, and climate change—are much more apparent in the southern half of our basin than in the upstream northern half (Danz et al. 2007).

Aboriginal peoples, who have lived here since time immemorial but with greatly diminished numbers during recent centuries, tend sacred flames in some locales that are not as severely abused as is the rest of our basin. Through the previous half millennium of gross change in our basin, some ancient species like the lake sturgeon and common loon have survived, in decreased numbers. Five centuries from now, will these species still be here?

Overall, the cumulative effects of degrading cultural stresses (see box 2) in this vast ecosystem may have peaked or plateaued about 1950, at least in southerly waters. Gradually the overall intensities of some abuses have been reduced, but legacy effects from the settlement era (Trautman 1977) remain 200 years later. Remediation of some old stresses may have offset additional new stresses since 1950. Perhaps the present century will bring with it major net recovery and rehabilitation, but it seems highly unlikely to us that Lake Erie, for example, will again come to resemble closely its ecosystemic state in the year 1500.

With precautionary commitment to sustainability, the inevitable new stresses may not cripple desirable ecosystem features to the extent that occurred with classes of old stresses over previous centuries. But sooner or later, a very human technician may do something foolish that will cause a nuclear electricity-generating plant to malfunction and blast radionuclides far and wide. Another 500 years of economic and population growth do not bode well for some of the remaining ancient features of these ecosystems, given past trends (Naidoo and Adamowicz 2001; Clausen and York 2008), especially with respect to freshwater fish—and including lake sturgeon (Miller Reed and Czech 2005; Rose 2005; Leprieur et al. 2008). Those past trends in economic and population growth have occurred even longer along the Atlantic coast of North America and in Europe. And the European sturgeon *Acipenser sturio* has been extirpated from the Seine Basin (Oberdorff and Hughes 1992) and most of Europe, whereas Atlantic sturgeon *A. oxyrinchus*, shortnose sturgeon *A. brevirostrum*, and Alabama sturgeon *Scaphirynchus suttkusi* have been listed as endangered in the United States (Jelks et al. 2008). As a result of overexploitation, habitat fragmentation, and water pollution, the International Union for Conservation of Nature considers sturgeon the most threatened group of animals, with 85 percent of the species at risk of extinction (http://www.iucn.org/?4928).

Our approach here may be termed "heuristic" and consistent, we suggest, with the "Prescription for Great Lakes Ecosystem Protection and Restoration" (Bails et al. 2005). We emphasize general concepts and inferences and do not mobilize in a quantitative and critically convincing way much of the empirical information in hundreds of references as might be expected in a state-of-the-art meta-analysis. Other chapters in this volume contain much relevant empirical information. We suggest that all the concepts and data combined critically may demonstrate that an appropriate measure of the ecological health of rehabilitating populations of the lake sturgeon species can serve as an integrative indicator of the state of ecological health of the aquatic part of our basin ecosystem. Who better to act as guardians of recovering sturgeon populations than their friends since time immemorial, the Aboriginal peoples?

Ancient Eco-History

Ancestral fish that resemble our present lake sturgeon have thrived in our Laurentian Basin and nearby watersheds for millions of years. Aboriginal humans, treading lightly in the lands and floating smoothly on the waters, "settled" the basin ecosystem many millennia ago, to be followed some centuries ago by many Europeans, with heavy tread and erosive navigation, followed in turn by many less-abusive humans of

Box 2. Terms for Cause-Effect Sequences in Complex Ecosystems

As we learn more about life in ecosystems, including microbial life involving vast numbers of microbes not yet identified specifically except through a few genes in their genomes, we now realize that full scientific understanding of any dynamic ecosystem is far beyond human capabilities. So the particular effects on an ecosystem of particular actions by humans cannot be fully inferred. The terms "cause" and "effect" may seem unambiguous, but, applied to the consequences of human actions on ecosystems, they cannot be made fully operational.

Numerous sets of terms have been used by ecosystem researchers in reference to cause-and-effect sequences, for example, action and reaction; stimulus and response; stress and strain (in physical sciences); stress and stressed; stressor and stress (in biological and psychological sciences); pressure, state, and response; and drivers, pressures, state, impact, and response.

Note that in the physical sciences "stress" usually refers to a cause. In biological and psychological sciences influenced by the 1960s works of Hans Selye, "stress" refers to an effect of some relevant cause or "stressor." Years after Selye introduced his terminology he came to realize that it was logically inconsistent with conventional uses in the physical sciences, but by then Selye decided not to change his terminology to align it with that of the physical sciences. In our title for this chapter we have used the term "stressed" for an effect of a causal "stress," that is, "stressed" implies a "strain." We continue with this convention throughout the chapter.

In diagnostic and forensic sciences a researcher may try to deduce or induce causes from observed effects. With ecosystems and other complex phenomena,

African and Asian origins only some decades ago. Each of these waves of immigrants came to interact with the Basin ecosystem in somewhat different ways. The most pervasive, prolonged and violent interactions between humans and other parts of these ecosystems came and stayed with the European invaders acting in what they thought were enlightened ways. The story of how the complex of lake sturgeon populations has fared through history in our basin's waters is introduced in this chapter, starting billions of years ago.

Here the waters of the "Laurentian Basin" are defined as those that now drain naturally into the string of large "lakes" (Superior, Huron-Michigan, Erie, Ontario) and large "rivers" (St. Mary's, St. Clair, Detroit, Niagara, St. Lawrence, Ottawa, Champlain) that eventually flow into the sea northeast of Quebec City, with tidewater inferring reliably about causality is difficult for many reasons. For example, the dynamics linking causes and effects may not be quantitatively linear but instead may involve nonlinear state shifts that resemble qualitative rather than quantitative consequences. Also, an ecosystem may respond to a variety of stimuli in ways that seem quite similar to an inexpert observer. Thus "eutrophication" is commonly used as a catchall term for a set of somewhat similar consequences, that is, a general adaptive syndrome, to a variety of quite different human actions that affect an aquatic ecosystem. Teasing out the particular cause(s) of an "eutrophication event," that is, from a general adaptive syndrome, with diagnostic or forensic scientific methods is difficult (Rapport et al. 1985).

That said, the art and science of ecological assessments have improved greatly in recent years with increased availability of large, consistently collected databases and sampling from hundreds of sites at continental or subcontinental scales (e.g., Pont et al. 2006; Brazner et al. 2007; Paulsen et al. 2008). In those studies, spatially extensive stresses such as row crop agriculture, urban development, hydrological alterations, and mining were determined to limit biological indicators from algae to fish. Paulsen et al. (2008), using a relative risk assessment approach developed by Van Sickle and Paulsen (2008), reported that for the coterminous United States excess total phosphorus, excess total nitrogen, and excess fine streambed sediments most often limited macroinvertebrate assemblages in wadeable streams—at least among the limited number of stresses quantitatively examined. These same stresses likely limit biota in lakes as well.

as an approximate downstream boundary. Important structural features of these waters' geological basin reflect ancient faulting in the underlying core or craton of our continent, which was first formed four or more billion years ago. Thus the wide "valley" that extends in an approximately straight line from Chicago northeastward past Quebec City and beyond parallels an ancient fault, to which other faults at right angles to it are joined in various locations.

The faults, as deep geological features of our part of the craton, are still active, as demonstrated by occasional earthquakes of low intensity (Dineva, Eaton, and Mereu 2004). The faults have been complemented at the surface by effects of numerous continental glaciers that arose periodically to the north of our basin and gradually flowed southward through it, approximately at a right angle to the natural flow of these waters when the glaciers were not present. Lakes were carved out and deepened; and the land's surface was scoured away. The scraped-off clay, sand, gravel, and rocks were deposited in massive humpy moraines at the southernmost edge of a particular glacier's motion. Many earlier moraines were overrun and deformed by later glaciers that proceeded further southward.

Glacial moraine complexes are of great importance to the aquatic ecosystems of the Laurentian Basin. Because of their porous structure, the moraines act like sponges that are filled during heavy rains of spring and fall and then drain continuously during the drier winter and summer months. Mining the moraines for sand and gravel diminishes their capacities to store and then deliver water into headwater streams, so flows of many streams during dry periods must now be less than was the case five centuries ago (e.g., Carlisle, Wolock, and Meador 2011). Perhaps some streams that once served as spawning habitat for sturgeon no longer do so now for this reason alone.

Springs that originate in moraine aquifers have relatively constant temperatures over a year that are about 1 to 2°C above average annual air temperature at the moraine site; snow cover acts as an insulating blanket in winter to keep the annual mean groundwater temperature a bit above the annual mean air temperature. So moraine springs average about 1°C at the northern edge near Lake Nipigon to about 12°C at the southern edge south of Cleveland (Schlesinger and Regier 1982). The drainage of the insulated water from moraines keeps the streams relatively warm and flowing in winter and relatively cool and flowing in summer, to the advantage particularly of certain fish species and other aquatic taxa. From a limnological perspective, the waters from the southerly moraines echo some of the seasonal features of the waters from the northerly part of our basin, with its porous ancient rock. Some species, like brook trout, appear in streams farther south latitudinally than would be the case in the absence of such moraines to provide habitat of suitable temperature year round (see, e.g., Moerke and Lamberti 2006).

During the past million years, these waters with their biota and especially the fish of the Laurentian Basin's aquatic ecosystem have been pushed south repeatedly by advancing continental glaciers into the headwaters of more southerly basins and have shifted north again as the glaciers melted thousands of years later. The biota of this relatively open aquatic ecosystem, as a whole or in several parts, likely always maintained some self-organizational ecosystemic integrity as it/they advanced southward in front of the ice and returned northward as the ice melted. Such integrity likely included complex structures such as "holonic" nesting of smaller ecosystems within larger ecosystems through a number of self-organized levels. The term "holonic" refers to a more generalized notion of vertical and horizontal nesting with reciprocal interactions than is usually implied by the etymologically misleading term "hierarchic" or "priestly order" (Regier 2008). At any particular time, spatial boundaries, usually indistinct, of ecosystems at various levels of nesting were likely related more clearly to hydrostatic gradients and hydrorheic features than to solid hydrographic features.

If photographs, say from an observation platform in space, could have been taken annually at midsummer over the past million years, the photos would likely show this sequence:

- At times between ice ages as now, an aquatic riverine treelike network with numerous vertical and lateral bulges (i.e., "lakes") of various sizes and with combined outflow northeastward to the Atlantic
- Early in an ice age, a distorted aquatic system with its outflow to the northeast dammed by ice twisting and shifting southward with advancing glaciation that broke into a number of subbasins with outflows southward during that ice age
- Late in an ice age, the hydrologically adapting aquatic system shifting northward and eventually settling into a configuration somewhat similar to that of the previous ice-free period when the ice dam in the northeast had melted

The natural aquatic ecosystem of five centuries ago, before the invasion by Europeans, had been affected by numerous north-south oscillations, with its separation into different southerly watersheds or catchments followed by a reverse shift into one watershed, as at present. From such a geological perspective, one would expect that the biota of this aquatic ecosystem would have a strongly riverine character. The native fish of our basin demonstrate evolution and adaptation to a metariverine habitat. One of the key considerations is that few of our basin's native fish species were thoroughly adapted to an offshore, pelagic lifestyle that included midwater spawning. Almost all spawned in rivers and tributaries or on reefs in the "lakes" across which currents flowed. The lake sturgeon was a part of such riverine wanderings for countless millennia.

Note that the Great Lakes are often said to be "young" geologically. The present/ recent hydrological/hydrographical manifestation is less than 10,000 years old. But many biotic parts of these ecosystems—the biota self-integrated into flexible, adaptive, riverine ecosystems—are much older than that. This becomes apparent when considering the complex stock structures of the salmonid and percid families of fish, say. So ecologically the biotic part of our basin ecosystem was not "young" five centuries ago; it was predominantly old-growth riverine, which has been mistaken by some for youth.

A complementary error has been common: the "artificial eutrophication" that followed massive loadings of plant nutrients a century ago was said to have triggered "premature aging" of the relevant aquatic ecosystem when the effect could perhaps more realistically be caricatured as induced obesity through forced feeding of the pelagic part of the ecosystem, or "pelagification" (Regier and Kay 1996; Kay and Regier 1999; Dobson, Regier, and Taylor 2002). In aquatic ecosystems like those of our basin, the normal self-organizational "succession" over centuries leads to dominance by the benthic association of species, in a process that may be termed "benthification."

Our Basin's fish thrive in heterothermic environments and cannot control their body temperatures (cold-blooded) except by purposely selecting habitat of preferred temperature, if available. For such fish, the temperature of various parts of their aquatic habitats is a key determinant for the spatiotemporal organization of various features of life histories. Many of the basin's fish species have rather narrow, and different, ranges of temperature that are optimal for spawning or for growth, say. Not far beyond an optimal temperature for growth lies a lethal upper temperature, which may be less that 8°C above the growth optimum (Regier et al. 1996).

During ice ages, the presence of ice in some waters during summer likely permitted the salmonid family to thrive there. Members of this family now thrive in or near those parts of the basin's waters that are coldest in summer, such as moraine streams to the south, deep lake waters saturated with dissolved oxygen throughout the basin, and unpolluted surface waters to the north. This family includes salmonines like char, trout, and salmon, thymallines like the grayling, and coregonines like the whitefish, cisco, lake herring, and chubs.

Farther downstream from glaciers and moraines and in surface waters separated from hydrologically isolated deep cold waters, the water warms in summer to a temperature suboptimal for salmonids. Cool-water percids thrive in such waters, as well as other families, including in particular the acipenserid, the lake sturgeon.

Farther downstream from the sources of cold moraine water in summer, and in shallow waters protected from coastal upwelling of cold bottom water and from long-shore currents of cool water, the warm-water family of centrarchid sunfish and black bass thrive together with other families.

To emphasize, we expect that a spectrum of fish families—with cold, cool, and warm habitat preferences in summer—has existed in the basin ecosystem during ice-free ages and in the parts of the displaced ecosystem during ice ages for millions of years. During the past two centuries, the amount of habitat for the cold-adapted fish has shrunk through warming of many streams and reductions of oxygen concentrations in some deep waters. For example, the grayling was eliminated from Michigan following land clearing in the early 1900s (Richards 1976). Similarly, the warm habitat has shrunk in many southerly locales because of destruction of the shallow land-water ecotone. Cool-water habitat may not have been diminished as much as the other two types. Lake sturgeon are at home in the cool waters, in company with the walleye and yellow perch. These two percid species can cope behaviorally both in benthic-dominated and in pelagic-dominated ecosystems, but the sturgeon cannot thrive in midwaters because of its heavy bony features.

According to the "Ontario Freshwater Fishes Life History Data Base" (http://ecometric.ca/fishdb/main.htm), based to a major extent on the work of Scott and Crossman (1973), the preferred habitat of lake sturgeon is the bottom water of lakes and large rivers 5 to 10 meters deep, where it feeds on benthic invertebrates in clay, mud, sand, and gravel bottom. It spawns in rivers in spring with water temperatures between 9 and 18°C; its estimated temperature preference in summer is 15 to 17°C, though field data suggest that the sturgeon may actually prefer a range of temperatures a couple of degrees higher than that. With that range of temperature preferences, during summer the lake sturgeon leaves warm stream, river, and epilimnetic "lake" waters, at least in the more southerly parts of the basin, and moves downslope within lakes to cooler waters but not into the cold hypolimnetic waters. Retrospectively, this species would likely have found appropriate habitat during any stage of glaciation and deglaciation of our basin in the past million years.

Numerical measures of many indicators of the state of health of our basin ecosystem are used now and more are being proposed (e.g., Niemi and Kelly 2007). Various fish species can be used as integrative indicators of ecosystem health with respect to those parts of the ecosystem that they inhabit. The lake trout as a representative salmonid has been used in this way for decades in our basin with respect to large cold waters (Regier 1992); to a lesser extent the walleye is used with respect to cool waters and smallmouth bass with warm waters.

It may be timely to document the case for using lake sturgeon as another integrative indicator of ecosystem health for cool waters in summer; much of the relevant information is contained in other chapters of this volume. The implicit rationale that was used by a number of agencies to select the lake trout as an integrative indicator of ecosystem health (Regier 1992) can be used with lake sturgeon.

Riverine Features of the Near-Pristine Waters in the Year 1500 and Subsequent Changes

Rivers normally wander across a relatively flat landscape over the ages, even in the absence of glaciers. Even during glacial periods, the water in the form of deep ice extending a thousand meters skyward flows, if only slowly, but it does flow, as do streams in and under the ice. In the preceding section, the whole water mass of our basin is described as having river-like features, but with massive vertical and horizontal bulges in places we call lakes or glaciers. Internal to these bulges, whether downward as in lakes or upward as in glaciers, there are currents with some river-like features.

In our basin, some nonecological experts of decades ago routinely referred to large rivers like the St. Clair and Detroit as "connecting channels" between lakes. Implicitly this terminology depreciates the importance of these rivers with respect to the lakes that they "connect." To us it makes more sense to emphasize the riverine features and flows of the lakes and treat the lakes as "connecting bulges" or large pools (with some riverine features) between the natural rivers. (This is also similar to the way some limnologists view the Amazon Basin and its lakes, slowly flowing swamp streams, and flooded forests.)

The tributaries from headwaters to one of the large lakes are obviously riverine. Especially in its northerly half, our basin has thousands of small lakes that rest on top of our billions-of-years-old craton, the Laurentian Shield. It also has numerous medium-sized lakes such as Fox, Nipigon, St. Clair, Nipissing, Simcoe, Kawarthas, Finger Lakes, Oneida, Opeongo, Champlain, and so on. There are four large lakes: Superior, Huron-Michigan, Erie, and Ontario. Five centuries ago, all of these lakes, whether of small, medium or large size, manifested notable riverine features, some of which have now been debased into "reservoirine" features (see box 1).

An important feature of natural waters in our part of the continent is that they are bordered by a land-water ecotone or riparian zone profusely vegetated in locales other than wave-swept, rocky shorelines (Minshall et al. 1985; Gregory et al. 1991; Sedell et al. 1991).

Five centuries ago, woody plants were found bordering shores often in dense thickets; woody debris of all sizes was found in headwater streams and down the tributaries along the shores and on the bottoms of rivers and lakes of increasing size all the way downstream to the St. Lawrence River. From what we now know about woody debris in relatively natural waters such as the Amazon Basin of South America and relatively pristine watersheds elsewhere in North America (Gregory, Boyer, and Gurnell 2003; Sedell and Froggatt 1984; Triska 1984; Hughes, Rinne, and Calamusso 2005a; Benke and Cushing 2005), we infer that woody debris had major ecosystemic importance in the pristine waters of our basin.

In the downstream half of our basin, much "wood" was already cut and cleared out of the terrestrial and aquatic parts two centuries ago. But some of the cut wood was then dumped into the waters as bark and sawdust that took decades to decompose anoxically; old sawdust and bark still remains in many inland lakes near old lumber mill sites (R. M. Hughes, personal communication). Some of the second-growth deciduous trees that followed thorough removal of the white pine forests then toppled into naturally eroding rivers to good ecological effect, but in recent decades these have been removed from popular canoeing rivers to facilitate passage. Natural freshwater ecosystems in forested regions like our basin need constant supplies of foliage from living woody plants and of dead wood (though not sawdust) to behave naturally, so our basin's aquatic ecosystems have become progressively pauperized for this reason, as well as others.

In their natural state, beaver in countless locales dammed smaller rivers. We infer that the beaver did not "engineer" their dams to withstand an intensity of natural flooding that occurred less frequently than about once in a decade or two, maybe. So these dams were generally topped by high flows and freshets and were breached repeatedly. Beaver preferred early-successional trees like aspen, which thrive on old beaver meadows, so their engineering design may have incorporated "planned obsolescence." In any case, beaver apparently did not permanently separate parts of streams behind permanent dams. Starting a century before the European agricultural settlers came, the demand for beaver pelts by the fur traders led to the great reduction in beaver and consequently of intact beaver dams.

In thousands of streams, of all sizes from headwater brooks to the St. Lawrence River, artificial dams (of earthen, timber, rock, concrete, and steel materials) were built for many purposes by the European invaders. Many of the first-generation dams, built with less expertise than were the beaver dams, were quickly washed out by spates that had been made more "flashy" and violent by clearing the woody plants and debris upstream, but a fifth-generation dam some decades later may have then persisted for many decades. A dam that was not breached or overtopped by annual floods became a permanent barrier to migrant fish spawners that may habitually have homed to spawning areas upstream of such a dam. In aggregate, this happened with respect to thousands of gorges, rapids, and reaches in streams and rivers.

Smith (1995) summarized some numerical data on the historical changes that occurred as follows: "the forests of the Lake Ontario drainage were described as 'unbroken' in 1784.... The process of forest cutting progressed from east to west; by the 1890s, the drainage area of Lake Ontario in western New York was described ... as 'almost entirely deforested.' Concurrent with the removal of forests was the drainage of swamps to create farmland and the construction of dams for water-powered mills. The construction of mills to process forest and farm products started in the late 1700s. By 1845, there were 7,406 sawmills run by water power in the state of New York ... and a somewhat lesser number of grist mills, plaster mills, tanneries, and other water-powered industries on both the United States and Canadian sides of Lake Ontario." (Note that the catchments for all these streams are not stated explicitly to have been parts of the Lake Ontario catchment.)

Large permanent dams also led to nonnatural, artificial flow regimes. Some changes fostered emergence of "dead zones" of anoxic mud and stagnant bottom water upstream of dams and in offshore deep waters in lakes.

Cropping away of smaller streams as such throughout larger catchment areas in the southern half of our Laurentian Basin has occurred progressively for centuries. Tile drainage and adaptation of storm sewers to conduct streams underground are common in urban and agricultural areas (Smith 1971). Examination of a historic series of detailed topographical maps shows that many streams of low order—first, second, even third—do not appear as surface streams in later maps (Steedman 1986). For more recent examples of this phenomenon in the conterminous United States, see Stoddard et al. (2005) and Carlisle, Wolock, and Meador (2011).

Gottgens and Evans (2007, 87) state that 632 dams are or were located on Ohio tributaries to Lake Erie. Of these, "29% are considered to be 'highly hazardous' and an additional 30% are considered to be 'significantly hazardous.' . . . Over time, hydraulic structures such as dams and reservoirs gradually change from assets to liabilities that cannot be ignored." Presumably the assets were perceived as such from a conventional economic perspective and not necessarily from an ecological perspective.

Large swampy areas were drained over a period of two centuries, such as the vast wet forest that extended from around the southwestern end of Lake Erie, northward around Lake St. Clair, and further northward to Saginaw Bay. Long stretches of the shores of larger rivers and lakes were diked. Some stretches of shorelines, especially those adjacent to dredged and deepened mouths of these rivers where they joined the large lakes and rivers, were sculpted and then hardened with concrete and steel to create harbors. The draining, diking, dredging, and hardening separately and together lopped off large ecotone areas of the natural riverine ecosystem, especially highly productive lake estuaries and floodplains.

A different sort of alteration has occurred where Lake Huron drains into the St. Clair River. In 1859, a 4–5 m deep rapids existed there. To facilitate commercial navigation, channels were cut and blasted through the bedrock in the 1920s (7 m), 1930s (8 m), and 1960s (9 m). The river is now continuously eroding the softer substrate and currently ranges from 10–21 m deep, resulting in a lowering of Lake Huron and Lake Michigan by 40 cm to date. This amounts to a vast volume of water when multiplied by the lakes' areas, and results in receding shorelines. Other changes to facilitate shipping (Erie Canal, Welland Canal, Soo [Sault] Locks, St. Lawrence Seaway) have opened up the Great Lakes to innumerable invasive alien species brought here in ships from throughout the world.

Another ecologically important riverine feature of our large "lakes" is that currents of water flowing through a lake or circulating in gyres within a lake scrub up against the shore and bottom (Rao and Schwab 2007). The Coriolis force, as related to the forward momentum of water that is flowing downslope because of gravity but deflected to the right in our Northern Hemisphere because of the rotation of the earth, is one reason for the curving currents in our rivers and riverine lakes. Winds, which are markedly changeable in direction in our basin, alter direction of flows and contribute to the strongly episodic features of the currents. Another reason relates to rocky barriers that deflect and channel forward motion of water. A shore where currents often flow or winds persistently blow across a long fetch, though not necessarily constantly, has riverine features with respect to bottom materials, rooted aquatics, woody debris, the benthos, and the fish assemblage. It is like a one-sided, episodic river that may be spatially continuous for a long stretch of coastline when a large gyre is rotating or in inshore waters in spring when a "thermal bar" is present (Rao and Schwab 2007).

The water of medium and large lakes shows one or more gyres, and sometimes gyres within gyres, of varying shape and velocity depending on seasonal meteorological and hydrological conditions. These have some features of slowly rotating whirlpools in rivers.

Many kinds of "developments" that European invaders created in our basin have caused changes in the current regimes in rivers and lakes, especially in the southerly downstream half of the basin. Perhaps one of the "strains" in the rheic system caused by these stressful developments was that the current patterns have become much coarser and more grossly episodic (see box 2). If so, such a change must have been disadvantageous to many species, including lake sturgeon.

Lake sturgeon prefer bottom habitats with currents; in such places the bottom may be of stone cobble and hard clay. For example, the underwater delta and distributary currents of the relatively warm Niagara River water into the colder Lake Ontario waters were a favorite resting, feeding, and spawning locale for sturgeon with a thriving fishery in the late nineteenth century. Rao and Schwab (2007) provide a description of the currents in this locale that show what must have been important features of the sturgeon's habitat preferences. The Niagara River delta was fouled with raw sewage from upstream cities, poisoned by pesticides partly as a result of massive spraying of Buffalo's elms with DDT against the Dutch elm bark beetle in the 1950s, and contaminated by the chemical industries along the shores upstream.

Similar physical and chemical alterations have been documented to affect lake sturgeon elsewhere. Elimination of lake sturgeon spawning habitats and innumerable low-head dams were associated with extirpation of lake sturgeon in the Red River of the North (Aadland et al. 2005). Freeman et al. (2005) described how migration barriers have resulted in the listing of lake sturgeon as threatened in the Alabama River system. Habitat fragmentation has decimated lake sturgeon numbers in the Wabash River (Gammon 2005) and the Wisconsin River (Lyons 2005).

One of the key sets of dynamic macro-features in pristine aquatic ecosystems in our basin was the incidence of massive episodic rheic features including river floods, shoreline storms, surface seiches, and thermal bars. When and where these were relatively predictable and gradual—on annual, decadal, and century-long scales—they were part of the natural context for healthy adapted ecosystems. On balance, what invading European settlers did to the streams and adjacent lands predisposed them to more violent consequences of floods, storms, and seiches, and they became more erratic, "peaky," and erosive than was desirable for the kinds of stream biota that had adapted to the previous less violent and more gradual episodic regimes. Such changes have been particularly obvious in rivers in urbanized, surface mined, and overcultivated catchment areas (Dodge 1989).

Hydrological innocents among the European invaders tried to counter the bad consequences of damming, diking, channelizing, hardening of shores, and other changes on land by additional and more massive earthen, concrete, and steel confining structures. Such obstructions in turn led to an increase in frequency and intensity of particularly destructive environmental events further downstream. More steel and concrete was often added, and the waters even farther downstream adapted again with increased violence, unpredictability, flashiness, and frequency of the most extreme manifestations in this pathological feedback syndrome. By the year 1970, the steel-and-concrete syndrome interacting with the channelize-dam-drain-and-dike syndrome had helped to transform large parts of our natural riverine ecosystem to a type dominated increasingly by an artificial "reservoirine" syndrome in form and behavior. This artificiality has been most prominent the further south one looks in the basin. But efforts to reverse this history are under way throughout our basin, mostly at a local level with removal of small dams that now serve no useful economic purpose (Evans and Gottgens 2007).

One ecological example of the effects of the transformation of our waters from riverine to reservoirine is the fish-related part of this deep ecosystemic transformation. The 200 or so fish species of the near-pristine state were generally adapted to benthic riverine features in these waters and to shallow waters of the riparian ecotone. Currently the dominant fish species are aliens preadapted to enriched pelagic lacustrine (or marine continental-shelf waters) that experience less difficulty in our "reservoirine waters" than do our native species. Again, other human abuses have contributed importantly to this transformation of the fish assemblages from riverine to "reservoirine" (see box 1).

Ecological Effects of Fishing on Lake Sturgeon

In ecosystems like those in our Laurentian Basin, the ecological effects of conventional fishing in historic times need to be teased out from the effects of other human stresses on these ecosystems. In our Laurentian Basin, such analyses, building on current empirical studies, have been reported as far back as the 1850s, when George Perkins Marsh undertook an investigation for the Vermont government to discover why the brook trout fisheries were failing in its streams. More recently, a team effort used a "forensic ecosystem approach" to infer causes for the failure of walleye fisheries in western Lake Erie in the 1960s (Regier, Applegate, and Ryder 1969). This was a difficult task because—as we came to realize—some generalized symptoms or effects of stress were not specific to particular stresses. The complex of nonspecific, nondiagnostic symptoms together with some specific, diagnostic symptoms could be termed a "general adaptive syndrome, GAS, to ecosystemic stress" as analogous to the GAS that Hans Selye had inferred with respect to the physiology of stressed mammalian organisms (Rapport et al. 1985).

That 1969 walleye study in turn led to a Symposium on Salmonid Communities in Oligotrophic Lakes (Loftus and Regier 1972) in which severe effects of major stresses were teased apart. Again, a GAS was inferred; it involved suppression of the dominant "riverine" salmonid complex of species and emergence of a "reservoirine" complex of alien species, though we did not use the terms "riverine" and "reservoirine" at that time to summarize key features of these two types.

Particularly for lake sturgeon, everybody's favorite colleagues of years ago, W. Bev Scott and Ed J. Crossman (1973), summarized how European settlers had fished sturgeon nonsustainably to lead to deep population reductions already in the nineteenth century. In an exhaustive comparative study of historical changes in the fish assemblages in three bays of our basin, Whillans (1979) included information about lake sturgeon. He critically examined accounts of events from early settlement by Europeans in the eighteenth century to 1976 in Toronto and Burlington Bays of Lake Ontario and Inner Long Point Bay of Lake Erie:

- Records of lake sturgeon in Toronto Bay began in 1800 when it was not one of the more preferred species by those with refined tastes. Sturgeon used Toronto Bay only for brief periods in the spring, when they congregated in the estuary of the Don River. By the 1840s, sturgeon were reported to be declining at a time while fishing was not intense; deforestation, milling, and construction of dams likely stressed the sturgeon population.
- Records of sturgeon in Burlington Bay began in 1855. Transformations in this fish assemblage were first noted between 1859 and 1877; for example, the sturgeon population declined during this period. Between 1878 and 1892, sturgeon were fished heavily and their spawning grounds may have been harmed by water pollution.

- Records of sturgeon in Inner Long Point Bay began between 1880 and 1885. Between 1892 and 1910, numbers declined in association with increased fishing pressure; between 1924 and 1937 sturgeon numbers fluctuated.

Though Whillans noted numerous adverse practices on these bay ecosystems besides nonsustainable fishing, he inferred that the main cause of lake sturgeon decline in all three bays was selective commercial exploitation, which was not intense in waters in and near these bays until the 1870s.

Historically, under rapidly intensifying fisheries in the second half of the nineteenth century that were conducted in the traditional European way (preferentially removing the largest sturgeon available with relatively nonselective gear), the abundance of lake sturgeon in our basin waters apparently did not show evidence of greatly increased fluctuations in abundance from year to year. Their abundance, as reflected in primitive measures of catch per unit effort of fishing, decreased gradually from year to year rather than fluctuating wildly. Moderate fluctuations in catch from year to year may have been due more to differences in the weather that affected temporal and areal aspects of fish schooling and fisher access and success rather than to the overall availability of sturgeon. But some degree of fluctuation may sometimes have been caused by conventional exploitation (see Whillans 1979).

In our basin, very large sturgeon were occasionally caught decades after the commercial sturgeon fishery was catching only few sturgeon that were mostly small. Perhaps some sturgeon preferred habitats that were not fished intensely, with a consequence that something like sturgeon refuges may have existed and persisted somewhere in our basin's waters.

Venturelli, Shuter, and Murphy (2009, 919) conducted a meta-analysis of time-series data on the reproductive success of 25 species of exploited marine fishes as related to the size and age distributions of the spawning population. They cautiously inferred that "a population of older, larger individuals has a higher maximum reproductive rate than an equivalent population of younger, smaller individuals, and that this difference increases in the reproductive lifespan of the population. These findings (i) establish an empirical link between population age structure and reproductive rate that is consistent with strong effects of maternal quality on population dynamics and (ii) provide further evidence that extended age structure is essential to the sustainability of many exploited fish stocks."

The report by Venturelli, Shuter, and Murphy supports the informed judgment of various experts on fisheries in our Laurentian Basin for many decades past that ensuring the presence of numerous large spawners in a fish population should be part of a sustainable fisheries policy. Refuges from which fishing is excluded rigorously may be a particularly helpful approach concerning the conservation of older and larger fish such as lake sturgeon. The underwater delta of the Niagara River at its mouth in Lake Ontario could be included in such a refuge for lake sturgeon.

If additional research supports the work of these authors with respect to marine fisheries and also provides evidence that such inferences are valid with respect to freshwater fisheries and lake sturgeon in particular in our Laurentian Basin, then the regulatory and ethical codes of fishers should be revised to reflect those inferences (Dobson, Regier, and Taylor 2002).

Benthification, Pelagification, and Surficialization

Back to an ecosystemic general adaptive syndrome, GAS, adapted from the physiological GAS of Hans Selye: in our basin's waters a dominant feature of this GAS was "pelagification" in that the normally dominant benthic association was suppressed and a somewhat unnatural pelagic association was fostered (Kay and Regier 1999). Again, such "pelagification" has often been termed "eutrophication," a potentially misleading term as applied in our basin.

More recently there have been reviews of relevant information about the failure of fisheries in multistressed enclosed seas (Caddy and Regier 2002). Such an "ecosystemic forensic approach" may not yet have been applied to problems in the Gulf of St. Lawrence, downstream from our basin ecosystem, which has its own sturgeon species.

Studies like those mentioned above have generally led to inferences that the strains or adverse effects of numerous (but perhaps not all) human-related stresses that are typical of conventional economic and cultural development tend to interact synergistically in aquatic ecosystems to augment the adverse effects of each and thus to trigger a GAS as an ecosystemic phase or state shift. Apparently the effects of those other cultural stresses have not interacted with the effects of conventional fishing stress to counteract the deleterious effects of the latter. Instead, the negative effects of various stresses may more frequently have interacted synergistically with those of conventional fishing to exacerbate further the harm done to the native fish by such fishing.

The artificial pelagification of the more stressed bays and shallow subbasins in our lakes that reached peak intensities several decades ago has been followed in turn during the past two decades by a suppression of that artificial pelagic association through the combined "benthification" to a new kind of artificial benthic association (alien zebra and quagga mussels, round goby, etc.) plus "surficialization" to an unusual kind of surface association (blooms of toxic algae, floating mats of decaying algae, etc.). This latter surface-bottom combination, with little systemic organization in the intervening waters, has contributed in some way to the emergence of a different type of "dead zone" in some deep waters, now with the presence of toxic organisms both at the top and at the bottom of these waters. Intense research is currently under way to describe and explicate phenomena that seem strange, at least at the scale and tempo with which they are appearing.

Rationally reliable scientific methods for conducting meta-analyses in multi-stressed ecosystems are evolving gradually (Zwiers and Hegerl 2008). New generations of scientists in our basin will presumably come to use them, perhaps rather noisily at first. Humility in the face of uncertainty, as urged by Jasanoff (2007), has been a common, if not universal, trait within the invisible college of our basin's researchers. New concepts and methods have come to be adopted smoothly, on the whole.

With respect to other human-generated stresses than fishing, the transformation from a riverine to a reservoirine type of ecosystem must have been severely detrimental on balance for sturgeon. In particular, dams on rivers prevented sturgeon from ascending them to their ancestral spawning grounds. Also, emergence of oxygen-depleted bottom waters of reservoirine ecosystems (see below) likely served to exclude sturgeon from some of their ancestral habitat in summer and perhaps in winter.

Phosphorus, Nitrogen, Oxygen, and "Dead Zones"

In the aquatic ecosystems favored by lake sturgeon, some common chemicals to which humans are relatively insensitive should be at concentrations of only relatively few parts per million of the watery environment. In particular, this statement relates to dissolved oxygen and dissolved forms of some strongly oxidized, as well as some strongly reduced, "hydrogenized," but still biologically reactive compounds of phosphorus and nitrogen.

The role of biologically reactive phosphorus compounds, which dissociate in water to yield phosphate ions, came into clear focus in the late 1960s in our basin. Phosphates (and nitrates, see below) are taken up during photosynthesis by plants with oxygen produced as a waste product; concurrently, and at times when photosynthesis is not occurring, oxygen is taken up in respiration by most of the organisms that are abundant in natural ecosystems of our basin.

Natural ecosystems in pristine states in our basin have ways to limit the concentrations of particularly reactive phosphate ions, for example, by complexing them—and thus rendering them "refractory" or less reactive biologically—on organic particles that originate from the slow decomposition of woody debris, say. But such complexing can be overridden by reductions in dissolved oxygen due to bacterial respiration during decomposition of readily oxidized organic material such as sewage and dead plankton. For purposes of respiration when dissolved oxygen falls to low levels, some organisms can steal oxygen from the strongly oxidized phosphate ions and thus transform them into more soluble phosphite ions (which can subsequently be oxidized again into temporarily uncomplexed phosphate in well-oxygenated water and contribute to photosynthesis again, etc.). So a positive feedback loop can emerge, as at the bottom of deep basins within lakes and reservoirs, that can result in a large anoxic "dead zone" during the four or five summer months when the overlying waters are stratified by temperature that blocks access of atmospheric oxygen to the deeper zones.

In our basin, deep anoxic waters are colder (at 4–10°C, say) than those preferred by lake sturgeon (15–17°C or a couple of degrees warmer). But under unusual storm conditions, say involving a strong atmospheric front passing rapidly over a lake, waves may be induced at the surface and in the thermocline that separates cold and warm waters (Rao and Schwab 2007). Such a wave or seiche may result in anoxic waters sloshing into shallower, usually cool-water areas to the disadvantage of normal organisms that live there. Anoxic waters displaced by such an internal wave or seiche may harm strongly benthic lake sturgeon. In Lake Erie, for example, the band of water along the bottom that falls in the temperature range of 15–17°C in summer, as preferred by lake sturgeon, is not wide; thus sturgeon may sometimes have to choose between anoxic colder waters and oxic warmer waters rather than waters in their preferred temperature range. Such a seiche that induces them to greater mobility than usual to escape cold anoxic water may make them more vulnerable to capture in stationary fishing gear. Trapped sturgeon may then be "drowned" by the anoxic water, as has occurred with other species.

Of course, what is usually termed a "dead zone" is not devoid of life. Unless it is poisoned by some nonselective toxic chemical or biocide, such a zone teems with life. But these life-forms, some of which resemble our genetic ancestors of billions of years ago and some that live in our own gastrointestinal tracts, thrive in the absence of dissolved oxygen. A relevant report by P. J. Mulholland and 30 colleagues (2008) is titled "Stream Denitrification across Biomes and Its Response to Anthropogenic Nitrate Loading." This team undertook a carefully designed experiment that involved loading isotope-labeled nitrate compounds into 72 streams in eight regions of the coterminous United States. The experimental design was based on many previous, less-comprehensive studies.

The team distinguished how the denitrification process differed in streams of three size categories draining smallish catchment areas that were relatively natural or subjected to urban or agricultural regimes. The team also distinguished how those denitrification processes differed as influenced by different concentrations of the soluble nitrate ions in the nine experimental cells (three size categories by three kinds of predominant human activity).

In their 2008 report, the authors did not focus on historic changes in any stream, say as it progressed from a cold, permanently free-flowing first- or second-order stream to an underground conduit or an intermittent stream following urban or agricultural development. Their information could perhaps be interpreted to imply that such trimming away of first- and second-order streams made those waterways less effective in denitrifying excessive loads of nitrates that washed in from the overly enriched lands of a catchment area. They conclude:

> Our findings underscore the management imperative of controlling nitrogen loading to streams and protecting or restoring stream ecosystems to maintain or enhance their nitrogen removal functions. Controlling loading to streams and stream nitrogen export is a proven solution to eutrophication and hypoxia problems in downstream inland and coastal waters. Our findings suggest caution before implementing policies (for example, reliance on intensive agriculture for biofuels production) that may yield massive land conversions and higher nitrogen loads to streams. Associated increases in streamwater [nitrate] concentration may reduce the efficacy of streams as nitrogen sinks [through denitrification], yielding synergistic increases in downstream transport to estuaries and coastal oceans. (Mulholland et al. 2008, 204)

If each of the separate narratives of phosphates and nitrates in our basin ecosystems is already complex, the combined narrative is even more so. For starters, if phosphates are relatively more abundant than nitrates, the rapid photosynthesis and growth of aquatic plankton may demand more nitrates than are being brought into such waters through inflows. Surface blooms of specialized algae may then occur spontaneously that have the ability to transform gaseous nitrogen from the atmosphere into oxygenated nitrogen to be used in the rapid photosynthesis. Again, this may lead to a vicious cycle, or positive feedback, with more living and then dead plant material, more decomposition, less dissolved oxygen, regeneration of dissolved phosphates, more fixing of atmospheric nitrogen by surface algae, eruption of toxic algae, and so on. Altogether such a complex syndrome is dominated by nonlinear processes, including feedback and phase or state shifts, for which no conventional quantitative methods provide reliable predictions.

In Lake Erie a "dead zone" that appeared with increasing regularity and size some six or more decades ago was part of the ecosystem-wide pelagification transformation. It was largely remediated through various controls on phosphate loading; and the relatively rapid natural flushing of Lake Erie helped. But then about two decades ago a somewhat different kind of "dead zone" appeared in Lake Erie as well as strange ecosystemic functions in other areas of bottom sediment that were not within a massive "dead zone." Zebra and quagga mussels seemed to be playing strong roles in the strange behavior of Lake Erie's aquatic ecosystem, as concerns the dynamics of phosphates, nitrates, oxygen, and various biota. Poisonous substances were being generated in the benthos in unnaturally large amounts that killed fish and then killed the birds that ate the fish. Poisonous algal blooms appeared more frequently at the surface of these waters. The many researchers at work studying this issue have yet to come to a consensus on what is causing this unwanted complex phenomenon.

Presumably the ecosystem pathologies caused by excessive loadings of phosphates and nitrates into our basin waters, including complex effects that follow as the ecosystems try to adapt to these loadings with their synergistic feedback and interactive loops, do not make life more pleasant for lake sturgeon and most other native species of our basin. Unlike some alien species, lake sturgeon do not thrive in severely modified, reservoirine ecosystems.

Of equivalent importance to the phosphate-nitrogen-oxygen complex pathology sketched above is the issue related to chemicals that strongly affect the genetic, hormonal, immunological, and other information capabilities of many organisms, including fish, birds, and humans. Chemical concentrations with the phosphate-nitrate-oxygen complex sketched above are commonly measured in parts per million. However, concentrations of hazardous contaminants may be dangerous at levels of parts per billion or trillion.

Many pharmaceutical, cosmetic, and sanitizing substances find their way into our waters via the sewage system; adverse ecosystem effects from their use have been coming into focus. Because of the widespread reliance on chemical medication, many in the medical profession, including both clinicians and public health types, seem compromised on this issue. But some public health professionals are trying to come to terms with adverse effects of residual medications prescribed by clinical physicians. It seems unlikely that permanent presence of hundreds of such chemical contaminants in these waters makes life more pleasant for lake sturgeon.

Recent studies seem to imply that when some organisms that thrive in waters nearly saturated with oxygen are subjected to waters with depleted but not zero levels of oxygen concentrations, they may exhibit symptoms like those triggered by some chemical contaminants, for example, endocrine disruption. If so, then an anoxic "dead zone" may develop a surrounding penumbra of hazardous waters with depleted levels of oxygen. Do fish like sturgeon sense such oxygen diminution, and do they have behavioral capabilities to evade such waters?

Climate Change

A meta-analysis of empirical studies relevant to climate change by Rosenzweig and colleagues (2008) is titled "Attributing Physical and Biological Impacts to Anthropogenic Climate Change." The authors examined thousands of data series, each longer than a 20-year duration, from across the globe. Many of these series relate to phenology; for example, the timing of a species' life-history events in spring. For North America, the authors found 94 percent of the 405 observed physical changes (that were statistically significant) to be consistent with what the authors would have expected with climate warming. Similarly, they found 88 percent of the 579 biological changes (that were statistically significant) to be consistent with their understanding of effects of climate warming. If there had as yet been no effects from any climate change, then the expected percentages would have been 50 percent in both cases. The authors apparently found no scientific reports, appropriate to their meta-analysis, concerning temperature-related changes apparent in extant data series from the aquatic ecosystems in our basin.

Numerous researchers have used a variety of approaches to assess some likely effects of forecasted scenarios of climate change on fish in our basin (e.g., Magnuson et al. 1989; Everett et al. 1995). Ralph Pentland (personal communication) estimates that climate change has already lowered the level of the Great Lakes by 5 cm, with more lowering expected. In general, researchers have inferred that the "moderate" extent of climate change to be expected in our basin would, in itself, not likely create insurmountable difficulties for many of our native species. Perhaps more correctly: forecasted climate changes would not likely cause many serious difficulties in the absence of the adverse effects of other human-caused stresses such as transformation of these waters from a riverine to a reservoirine state; introductions of alien species preadapted to a warmer climate; emergence of anoxic and toxic bottom waters and toxic surface algal blooms; and so on. But it seems unlikely that any of the other human-induced changes would predispose these systems to respond favorably to climate change, at least with respect to native species like the sturgeon. There may well be some alien species already present that will thrive with warming due to climate change, and more (including human pathogens) that are preadapted to a warmer climate will likely invade, or be released into our basin.

Concluding Comments

Our basin's fish and fisheries "hit the wall" ecologically and economically about half a century ago, in the 1950s. Numerous remediative and rehabilitative measures have been undertaken since then. The record concerning corrective measures with exploitative fishing and loading of phosphate nutrients shows important improvements. The record with channel modifications, nitrates, contaminants and alien species is at best ambiguous, but generally poor. Climate change caused by continued economic and population growth will likely interact badly with remaining features of the old stresses as well as with new stresses for which current remediative and rehabilitative programs are clearly inadequate and precautionary methods are still weak (Regier et al. 1989).

The lake sturgeon has been important to human fishers for at least two thousand years, judging from work by archaeologists (as in Smith 2004) and historians (as in Thwaites 1899), both with respect to sturgeon in the Straits of Mackinac. A quick search of the Internet has yielded five bays, 13 rivers, and two creeks in our basin named after sturgeon, not to mention human settlements with that name. Somehow, several small sturgeon populations have survived the past two centuries of intense ecosystemic abuse in our basin.

During recent centuries, lake sturgeon populations were suppressed and some were extirpated in our basin's waters (Scott and Crossman 1973, and other authors). A thorough forensic scientific study to identify the specific cause(s) of an extirpation of a particular lake sturgeon population was never undertaken (though work by Whillans [1979] deserves honorable mention) and cannot now be done in a fully convincing way. In recent decades some populations that survived all the abuse but in low numbers have been protected and are slowly increasing in abundance. If a full recovery of a population may take several of their generations, each of which may be 10 years long with sturgeon, then we should not expect to encounter a rehabilitated sturgeon population in our basin for half a century or so, and then only if ecosystemic rehabilitation—now stalled–resumes energetically. Assuming that successful sturgeon reproduction eventually comes to be dominated by large females (Venturelli, Shuter, and Murphy 2009), as with other long-lived large species, then we may not see widespread rehabilitation for a century.

Similar declines and extirpations of sturgeons have occurred throughout their ranges as a result of reproductive failure associated with fisheries exploitation, water pollution, and migration barriers. Many sturgeon species are threatened or endangered (Williams et al. 1989; Oberdorff and Hughes 1992; Otel 2007; Jelks et al. 2008) with little hope for recovery in ecosystems that have suffered severe

assemblages in the Americas. J. N. Rinne, R. M. Hughes, and B. Calamusso, eds. American Fisheries Society Symposium 45. American Fisheries Society.

———. 2005b. Historical changes in large river fish assemblages of the Americas: A synthesis. *In*: Historical changes in large river fish assemblages in the Americas. J. N. Rinne, R. M. Hughes, and B. Calamusso, eds. American Fisheries Society Symposium 45. American Fisheries Society.

Jasanoff, S. 2007. Technologies of humility. Nature 450:33.

Jelks, H. J., S. J. Walsh, N. M. Burkhead, S. Contreras-Balderas, E. Diaz-Pardo, D. A. Hendrickson, J. Lyons, N. E. Mandrak, F. McCormick, J. S. Nelson, S. P. Platania, B. A. Porter, C. B. Renaud, J. J. Schmitter-Soto, E. B. Taylor, and M. L. Warren Jr. 2008. Conservation status of imperiled North American freshwater and diadromous fishes. Fisheries 33:372–386.

Juncosa, B. 2008. Suffocating seas: Climate change may be sparking new and bigger "dead zones." Scientific American 299(4): 20, 22.

Kay, J. J., and H. A. Regier. 1999. An ecosystemic two-phase attractor approach to Lake Erie's ecology. *In*: State of Lake Erie (SOLE)—past, present and future. M. Munawar, T. Edsall, and I. F. Munawar, eds. Ecovision World Monograph Series. Backhuys Publishers.

Leprieur, F., O. Beauchard, S. Blanchet, T. Oberdorff, and S. Brosse. 2008. Fish invasions in the world's river systems: When natural processes are blurred by human activities. Public Library of Science—Biology 6(2): e28. doi:10.1371/journal.pbio.0060028.

Loftus, K. H., and H. A. Regier, eds. 1972. Proceedings of the 1971 Symposium on Salmonid Communities in Oligotrophic Lakes. Journal of the Fisheries Research Board of Canada 29(6):611-986.

Lyons, J. 2005. Fish assemblage structure, composition and biotic integrity of the Wisconsin River. *In*: Historical Changes in Large River Fish Assemblages in the Americas. J. N. Rinne, R. M. Hughes, and B. Calamusso, eds. American Fisheries Society Symposium 45. American Fisheries Society.

Magnuson, J. J., D. K. Hill, H. A. Regier, J. A. Holmes, J. D. Meisner, and B. J. Shuter. 1989. Potential responses of Great Lakes fishes and their habitat to global climate warming. *In*: The potential effects of global climate change on the United States, Appendix E, Aquatic Resources. J. B. Smith and D. A. Tirpak, eds. EPA-230–05–89–055. U.S. Environmental Protection Agency.

Miller Reed, K., and B. Czech. 2005. Causes of fish endangerment in the U.S., or the structure of the American economy. Fisheries 30(7): 36–38.

Minshall, G. W., K. W. Cummins, R. C. Petersen, C. E. Cushing, D. A. Burns, J. R. Sedell, and R. L. Vannote. 1985. Developments in stream ecosystem theory. Canadian Journal of Fisheries and Aquatic Science 42:1045–1055.

Moerke, A. H., and G. A. Lamberti. 2006. Relationships between land use and stream ecosystems: A multistream assessment in southwestern Michigan. *In*: Landscape influences on stream habitats and biological assemblages. R. M. Hughes, L. Wang, and P. W. Seelbach, eds. American Fisheries Society Symposium 48. American Fisheries Society.

Mulholland, P. J., A. M. Helton, G. C. Poole, and 28 others. 2008. Stream denitrification across biomes and its response to anthropogenic nitrate loading. Nature 452:202–205.

Naidoo, R., and W. L. Adamowicz. 2001. Effects of economic prosperity on numbers of threatened species. Conservation Biology 15:1021–1029.

Niemi, G. J., and J. R. Kelly, eds. 2007. Coastal indicators. Journal of Great Lakes Research 33 (special issue 3): 1–318.

Oberdorff, T., and R. M. Hughes. 1992. Modification of an index of biotic integrity based on fish assemblages to characterize rivers of the Seine Basin, France. Hydrobiologia 228:117–130

Ontario Freshwater Fishes Life History Data Base. Http://ecometric.ca/fishdb/main.htm.

Otel, V. 2007. Atlasul pestilor: Din rezervatia biosferei Delta Dunarii. Institutul National de Cercetare-Desvoltare Delta Dunarii.

Paulsen, S. G., A. Mayio, D. V. Peck, J. L. Stoddard, E. Tarquinio, S. M. Holdsworth, J. Van Sickle, L. L. Yuan, C. P. Hawkins, A. T. Herlihy, P. R. Kaufmann, M. T. Barbour, D. P. Larsen, and A. R. Olsen. 2008. Condition of stream ecosystems in the US: an overview of the first national assessment. Journal of the North American Benthological Society 27:812–821.

Perkins, J. 2004. Confessions of an economic hit man. Penguin.

Pont, D., B. Hugueny, U. Beier, D. Goffaux, A. Melcher, R. Noble, C. M. Rogers, N. G. Roset, and S. Schmutz. 2006. Assessing river biotic condition at the continental scale: A European approach using functional metrics and fish assemblages. Journal of Applied Ecology 43:70–80.

Rao, R., and D. J. Schwab. 2007. Transport and mixing between coastal and offshore waters in the Great Lakes: A review. Journal of Great Lakes Research 33(1): 202–218.

Rapport, D. J., H. A. Regier, and T. C. Hutchinson. 1985. Ecosystem behavior under stress. American Naturalist 125:617–640.

Regier, H. A. 1992. Indicators of ecosystem integrity. *In*: Ecological indicators. D. McKenzie, D. E. Hyatt, and V. J. McDonald, eds. Elsevier Applied Science Publications.

———. 2008. Hierarchy and holonocracy. *In*: The ecosystem approach: Complexity, uncertainty, and managing for sustainability. D. Waltner-Toews, J. J. Kay, and N.-M. E. Lister, eds. Columbia University Press.

Regier, H. A., V. C. Applegate, and R. A. Ryder. 1969. Ecology and management of the walleye in western Lake Erie. Great Lakes Fishery Commission Technical Report 15.

Regier, H. A., and J. J. Kay. 1996. An heuristic model of transformations of the aquatic ecosystems of the Great Lakes–St. Lawrence River Basin. Journal of Aquatic Ecosystem Health 5:3–21.

Regier, H. A., P. Lin, K. K. Ing, and G. A. Wichert. 1996. Likely responses to climate change of fish associations in the Laurentian Great Lakes Basin: Concepts, methods and findings. Boreal Environment Research (Helsinki) 1:1–15.

Regier, H. A., and K. H. Loftus. 1972. Effects of fisheries exploitation on salmonid communities in oligotrophic lakes. Journal of the Fisheries Research Board of Canada 29:959–968.

Regier, H. A., R. L. Welcomme, R. J. Steedman, and H. F. Henderson. 1989. Rehabilitation of degraded river ecosystems. *In*: Proceedings of the International Large River Symposium (LARS). D. P. Dodge, ed. Special Publication of the Canadian Journal of Fisheries and Aquatic Sciences 106. Canada Department of Fisheries and Oceans.

Richards, J. S. 1976. Changes in fish species composition in the Au Sable River, Michigan from the 1920s to 1972. Transactions of the American Fisheries Society 105:32–40.

Rinne, J. N., R. M. Hughes, and B. Calamusso, eds. 2005. Historical Changes in Large River Fish Assemblages in the Americas. American Fisheries Society Symposium 45. American Fisheries Society.

Rose, A. 2005. Economic growth as a threat to fish conservation in Canada. Fisheries 30(8): 36–38.

Rosenzweig, C., D. Karoly, M. Vicarelli, P. Neofotis, Qigang Wu, G. Casassa, A. Menzel, T. L. Root, N. Estrella, B. Seguin, P. Tryjanowski, Chunzhen Liu, S. Rawlins, and A. Imeson.

2008. Attributing physical and biological impacts to anthropogenic climate change. Nature 453:353–357.

Schlesinger, D. A., and H. A. Regier. 1982. Climatic and morphoedaphic indices of fish yields from natural lakes. Transactions of the American Fisheries Society 111:141–150.

Scott, W. B., and E. J. Crossman. 1973. Freshwater fishes of Canada. Fisheries Research Board of Canada Bulletin 184.

Sedell, J. R., and J. L. Froggatt. 1984. Importance of streamside forests to large rivers: The isolation of the Willamette River, Oregon, USA, from its floodplain by snagging and streamside forest removal. Internationale Vereinigung für theoretische und angewandte Limnologie Verhandlungen 22:1828–1834.

Sedell, J. R., R. J. Steedman, H. A. Regier, and S. V. Gregory. 1991. Restoration of human impacted land-water ecotones. In: Ecotones: The role of landscape boundaries in the management and restoration of changing environments. M. M. Holland, P. G. Risser, and R. J. Naiman, eds. Chapman and Hall.

Smith, B. A. 2004. The gill net's "native country": The inland shore fishery in the northern Lake Michigan basin. In: An upper lakes archeological odyssey. W. A. Lovins and C. E. Cleland, eds. Wayne State University Press.

Smith, P. W. 1971. Illinois streams: A classification based on their fishes and an analysis of factors responsible for disappearance of native species. Illinois Natural History Survey, Biological Notes 76.

Smith, S. H. 1995. Early changes in the fish community of Lake Ontario. Great Lakes Fishery Commission Technical Report 60.

Sproule-Jones, M. 2002. The restoration of the Great Lakes. University of British Columbia Press.

Steedman, R. J. 1986. Historical streams of Toronto. Toronto Field Naturalist 382:14–18.

Stoddard, J. L., D. V. Peck, S. G. Paulsen, J. Van Sickle, C. P. Hawkins, A. T. Herlihy, R. M. Hughes, P. R. Kaufmann, D. P. Larsen, G. Lomnicky, A. R. Olsen, S. A. Peterson, P. L. Ringold, and T. R. Whittier. 2005. An ecological assessment of western streams and rivers. EPA 620/R-05/005, U.S. Environmental Protection Agency, Washington, DC.

Thwaites, R.G., ed. 1899. The Jesuit relations and allied documents: Travels and explorations of the Jesuit missionaries in New France 1610–1791. Vol. 55, 1670–1672. Burrows Brothers. Http://puffin.creighton.edu/jesuit/relations/relations_55.html.

Trautman, M. B. 1977. The Ohio country from 1750 to 1977: A naturalist's view. Ohio Biological Survey Biological Notes No. 10, Ohio State University.

Triska, F. J. 1984. Role of wood debris in modifying channel morphology and riparian areas of a large lowland river under pristine conditions: A case history. Internationale Vereinigung für theoretische und angewandte Limnologie Verhandlungen 22:1876–1892.

Van Sickle, J., and S. G. Paulsen. 2008. Assessing the attributable risks, relative risks, and regional extents of aquatic stressors. Journal of the North American Benthological Society 27:920–931.

Venturelli, P. A., B. J. Shuter, and C. A. Murphy. 2009. Evidence of harvest-induced maternal influences on the reproductive rates of fish populations. Proceedings of the Royal Society B 276:919–924.

Whillans, T. H. 1979. Historic transformations of fish communities in three Great Lakes bays. Journal of Great Lakes Research 5(2): 195–215.

Williams, J. E., J. E. Johnson, D. A. Hendrickson, S. Contreras-Balderas, J. D. Williams, M.

Navarro-Mendoza, D. E. McAllister, and J. E. Deacon. 1989. Fishes of North America endangered, threatened or of special concern: 1989. Fisheries 14(6): 2–20.

Zwiers, F., and G. Hegerl. 2008. Attributing cause and effect. Nature 453:296–297.

EDWARD A. BAKER AND NANCY AUER

Habitat, Foods, and Feeding

Habitat

As the name implies, lake sturgeon are commonly found in lakes, and because of their biology they occupy large lakes. The Great Lakes lie at the center of the species native range (Harkness and Dymond 1961) and these lakes once supported a large concentration of lake sturgeon. Lake sturgeon were in all likelihood one of the most abundant large-bodied fish species in the Great Lakes prior to extensive settlement of the region, with estimates of abundance exceeding 16 million fish in all Great Lakes combined. The lake sturgeon commercial fishery that developed in the Great Lakes provides an indication of historic lake sturgeon abundance. According to statistics compiled by the Great Lakes Fishery Commission, during the peak of the Great Lakes commercial lake sturgeon fishery in the late 1800s, an average of over 4 million pounds of lake sturgeon was harvested annually. The maximum harvest occurred in 1885, when over 8.6 million pounds were harvested (Baldwin et al. 1979). The abundance of lake sturgeon in each of the Great Lakes was closely tied to the habitat and productivity of the individual lake. Lake sturgeon harvest was greatest in Lake Erie and was lowest in Lake Superior. Lake Erie has a large basin and is a relatively shallow, warm, and productive lake. There are also numerous large productive rivers that feed into the lake. However, in Lake Superior abundance was much less than in the other lakes. Although it is the largest of the Great Lakes, Lake Superior is much deeper, colder, and less productive than the lower lakes. Estimated lake sturgeon abundance in 1885 was over 11 million fish in Lake Michigan but was only 870,000 in Lake Superior (Hay-Chmielewski and Whelan 1997). Much has changed in the Great Lakes since the late 1800s, including widespread alteration of habitat, and lake sturgeon abundance in the Great Lakes today is estimated to be less than 1 percent of the peak abundance of the 1800s (Hay-Chmielewski and Whelan 1997).

The native range of lake sturgeon includes the Great Lakes and St. Lawrence River, the Hudson Bay drainage of Canada, and the Mississippi River upstream of northern Mississippi (Scott and Crossman 1973). Lake sturgeon were historically abundant and found in many of the large rivers and lakes in all of these major drainages. However, the status of lake sturgeon throughout its native range is similar to that of the Great Lakes; its abundance is greatly reduced from the peak before extensive settlement. The demise of lake sturgeon throughout the species range can be attributed to excessive harvest, but habitat changes have also played an important role in the decline. In the Great Lakes, these habitat changes are also likely preventing lake sturgeon populations from rebounding.

Broadly defined, habitat includes the physical places occupied by an organism and can include the other organisms present with which a species interacts. In the case of lake sturgeon, the physical habitats occupied are varied and depend on season and the age of the fish. Most of a lake sturgeon's life is spent in lake or large river habitat feeding and growing, so they are associated with habitats that produce the organisms on which lake sturgeon feed. Lake sturgeon prey on benthic animals or those that live directly on or in the bottom substrates of lakes and rivers. The bottom substrates in relatively shallow waters of lakes are the most productive for benthic animals. As a result, lake sturgeon spend most of their time in relatively shallow water in lakes, typically less than 30.5 m (100 ft) deep and most often less than 9 m (30 ft) deep in the Great Lakes (Harkness and Dymond 1961).

Because fish are cold-blooded and their body temperature, and therefore metabolic processes, are dictated by the temperature of the surrounding water, most fish exhibit preferences for specific water temperatures that optimize their growth and development. The water temperature preferences of lake sturgeon are not as well understood as for other species. For example, several trout species are known to actively seek out cold water and largemouth bass actively seek warm water. Lake sturgeon water temperature preference for spawning is well known. They spawn when water temperatures reach about 12°C (54°F) and will continue until water temperature reaches 18°C (65°F) (Harkness and Dymond 1961; Kempinger 1988; Auer 1996a). However, little is known of water temperatures that adult lake sturgeon seek out during the extended period between spawning. Based on results from experimental hatchery rearing of lake sturgeon, it is likely that they seek relatively warm water. In an experimental hatchery setting, lake sturgeon growth was greatest when water temperatures were 15–18°C (60–64°F), and mortality was observed when water temperatures reached 22°C (72°F), suggesting this is near the upper lethal limit for lake sturgeon (Wehrly 1995; Diana, Webb, and Essington 2003).

However, adult lake sturgeon congregations have been found in Lake St. Clair at a water temperature of 25.6°C (78°F) (M. Thomas unpublished). In streamside hatcheries that circulate river water, age-zero lake sturgeon have continued to feed at water temperatures as high as 28°C (82°F) for a few hours midday (E. Baker unpublished). Radio telemetry studies of juvenile (ages 1–5) lake sturgeon suggest that they move from deeper cool water during the daytime to shallow water at night and back to cooler deep water during the day (Holtgren and Auer 2004). Presumably the fish are feeding in the shallower water at night and resting in the cooler water during daytime, when shallow water habitats may become too warm, or move into deep water in response to intense light.

Lake sturgeon roam shallow nearshore Great Lakes waters, feeding throughout the year, and are known to travel long distances in search of food. For example, Green Bay in Lake Michigan is shallow and productive. It is also known to have a relatively high concentration of lake sturgeon when compared to other areas in Lake Michigan. Recent research has demonstrated that the lake sturgeon found in Green Bay originated in rivers that empty directly into Green Bay (Menominee River, Peshtigo River, etc.) as well as rivers on the eastern shore of Lake Michigan, including the Manistee (Bott 2006). The productive waters of Green Bay appear to be an important feeding area for lake sturgeon from throughout the Lake Michigan basin, and sturgeon may migrate to Green Bay to feed and grow and then return to their natal rivers to reproduce. The same appears to be true of Saginaw Bay in Lake Huron, Lake St. Clair, and the bays at the western end of Lake Superior. Because lake sturgeon occupy relatively shallow nearshore waters of the Great Lakes and feed on the bottom, their migrations most likely follow the shoreline and do not cross the open waters of the lake. In large rivers where lake sturgeon may live their entire lives (e.g., Niagara River, St. Clair River), they also occupy the areas that are most productive for the benthic organisms they feed on and may travel long distances in search of food and suitable spawning habitat.

The only time that lake sturgeon abandon lake and river feeding habitat is when, as mature adults, they respond to the urge to spawn. Spawning occurs in the spring, and adults return to the rivers of their origin to spawn. Because lake sturgeon roam great distances in search of food, the return trip to spawning rivers for some can be very long. For example, tag return information from Lake Superior indicates that lake sturgeon that spawn in the Sturgeon River in Michigan's Upper Peninsula near Houghton travel as far east as Whitefish Bay near Sault Ste. Marie, Michigan, and as far west as Chequamegon Bay near Ashland, Wisconsin, while feeding in Lake Superior; distances in excess of 200 km (125 miles) (Auer 1999). Tag return information from Lakes Michigan and Huron has shown that lake sturgeon from Lake Winnebago, Wisconsin, that spawn in the Wolf River have traveled down the Fox River to Green Bay and continued on to Saginaw Bay near Saginaw, Michigan, in Lake Huron, a distance of over 600 miles (1,000 km).

As winter gives way to spring and adult lake sturgeon shift their focus to spawning, they begin what can be a long migration up rivers to spawning habitats. As lake sturgeon ascend rivers, they depend on deep holes for rest and concealment (Threader, Pope, and Schaap 1998). Prior to extensive settlement of the Great Lakes region, most, if not all, large rivers were used by spawning lake sturgeon, and spawning migrations would take the fish dozens to over 100 miles (160 km) upstream to reach suitable habitat. Now, in the Lake Superior drainage, lake sturgeon ascend the Sturgeon River to spawn near the base of Prickett Dam, a distance of 43 miles (69 km) (Auer 1996a). In the Wolf River, Wisconsin lake sturgeon travel upstream 155 miles (250 km) from Lake Winnebago to reach their spawning habitats (Kempinger 1988). In many rivers in Canada lake sturgeon migrate upstream until they reach an impassable barrier such as a natural waterfall and spawn at the base of the falls (Harkness and Dymond 1961).

Most river habitats used by spawning lake sturgeon can be best described as rapids characterized by high water velocity with clean substrates of large gravel, cobble, or even bedrock. An important component of spawning habitat is the presence of abundant interstitial spaces or gaps between the rocks where lake sturgeon eggs can settle and be concealed from predators. Water depth apparently has little influence on spawning habitat selection, as lake sturgeon have been observed spawning in the St. Clair and St. Lawrence rivers in water up to 60 feet (18 m) deep (LaHaye et al. 1992), while in smaller rivers like the Sturgeon River in Michigan's Upper Peninsula and the Black and Manistee rivers in the northern Lower Peninsula of Michigan, lake sturgeon have been observed spawning in water only two to three feet (1 m) deep (Chiotti et al. 2008).

Because depth does not appear to influence spawning site selection, the most important physical habitat cues for lake sturgeon spawning are apparently adequate water velocity and the presence of clean substrate. Reported water velocities at lake sturgeon spawning sites in rivers are in excess of 3 ft/s or 0.91 m/s (Threader, Pope, and Schaap 1998; Chiotti et al. 2008).

There are also persistent, albeit unconfirmed, reports of lake sturgeon spawning in nearshore lake habitats. In Lake Michigan, spawning activity has been reported on a reef near the St. Joseph River and near Ludington, Michigan, along the eastern shore of the lake. In Burt Lake, a large lake in the northern Lower Peninsula of Michigan, lake sturgeon spawning activity has been reported along the eastern shore of the lake in relatively shallow water with rocky substrate. In instances where lake spawning has been reported, the habitat characteristics are similar and consist of an active wave zone along the eastern shoreline with gravel or cobble substrates. Whether spawning in lake habitats is actually occurring or successful is unknown.

The presence of interstitial spaces in the spawning substrate is important for successful lake sturgeon spawning because they are broadcast spawners. Broadcast spawning fish do not construct nests for spawning, nor do they care for the eggs after spawning. Instead, lake sturgeon select a spot for spawning and simply release eggs and sperm into the current and the eggs are fertilized as they drift downstream with the current. Lake sturgeon eggs are negatively buoyant and sink to the substrate shortly after they are released by the female. When there are gaps in the substrate the eggs sink down into those spaces and adhere to the surfaces of the rocks, which protect them from large predators like suckers and crayfish and also shelter them from the main current. If spawning occurs where there are no gaps for the eggs to sink into, the eggs are more vulnerable to predators and can also be dislodged by the current and carried downstream to habitats that are unfavorable for incubation.

Lake sturgeon eggs develop rapidly and hatching occurs in as few as five days after spawning (Kempinger 1988). At hatching, lake sturgeon are considered larvae because they are not fully developed and still have a large yolk sac that sustains them for another several days. They remain in the spaces in the substrate until the yolk sac is used up and they are ready to begin feeding. This stage lasts about seven days and ends when the larval lake sturgeon swim up out of the substrate and drift downstream to nursery habitat (Kempinger 1998; Auer and Baker 2002; Smith and King 2005).

Little is known of the habitat requirements of lake sturgeon for the first few weeks after they hatch. They are less than 2.5 cm (1 in) long and darkly colored when they hatch and drift downstream at night, probably to avoid sight-feeding predators, and do not drift during daylight hours. Larval lake sturgeon drift has been recorded up to 64 km (40 miles) downstream from the spawning habitat (Auer and Baker 2002), but the daytime habitats used by larvae as they disperse downstream from the spawning site have not been documented. Eventually they stop drifting and begin feeding. By midsummer the fish are sandy-colored with dark mottling and blend in well with the sand and mixed sand/gravel substrates in the rivers where they feed and continue to grow.

In the Sturgeon River of Michigan researchers found newly hatched young on gravel and mixed sand/gravel substrates but never on pure sand substrates (Holtgren and Auer 2004), while in the Peshtigo River, Wisconsin researchers studying young

of resting habitat that lake sturgeon relied on as they ascended rivers in the spring spawning run. Sediments also likely altered the spawning habitats in rivers by filling in the interstitial gaps that lake sturgeon eggs depended on for successful incubation. The widespread logging undoubtedly raised the temperatures of rivers because the canopy of trees that shaded river channels was removed and allowed more direct sunlight to hit the river surface and warm the water.

The habitat change that has probably had the greatest long-term impact on lake sturgeon populations occurred following the peak of the logging era of 1870–1890. As settlement of the Great Lakes region expanded, many rivers that these fish used for spawning were dammed. Sturgeon species in general are noted for their long upriver migrations to find suitable spawning habitat (Auer 1996b). In the case of lake sturgeon these migrations may exceed 240 km (150 miles) (Kempinger 1988). The construction of dams, primarily hydropower dams that produce electricity, has eliminated access to habitat in most rivers where lake sturgeon previously spawned, blocking migrations and flooding the most suitable spawning habitat under the reservoirs created by the dams. In most cases the dams constructed on tributaries to the Great Lakes were built on the first high-gradient river reach upstream from the lake. These high-gradient (defined by the drop in elevation over distance) river reaches are the most suitable for hydropower dam construction because they provide the highest hydraulic head and therefore are optimal sites for hydroelectricity generation. Hydropower dams were typically built at the base of high-gradient reaches, resulting in flooding of the river upstream of the dam. Unfortunately for lake sturgeon (and many other species), these high-gradient river reaches are the habitats needed for spawning. Dams in rivers throughout the Great Lakes region prevent access to the most critical habitat lake sturgeon need to complete their life cycle. The Menominee River forms the border between northeast Wisconsin and Michigan's Upper Peninsula and is a good example of a river that once supported a large spawning run of lake sturgeon but has been severely impacted by dam construction. Prior to dam construction, lake sturgeon (as well as walleye, lake whitefish, lake trout, and other fish species) would migrate up the river in search of suitable spawning habitat. Lake sturgeon historically could migrate as far as Sturgeon Falls near Iron Mountain, Michigan, a distance of approximately 161 km (100 miles) (Thuemler 1985). However, the construction of six dams has fragmented the Menominee River system, and the first dam upstream from Lake Michigan is a mere 4 km (2.5 miles) from the lake.

This dam prevents fish from Lake Michigan accessing and using more than 156 km (97 miles) of previously used river spawning and rearing habitat. Barriers to migration are common to most of the rivers historically used by spawning lake sturgeon across the Great Lakes (Holey et al. 2000). In Michigan alone, 90 percent of the largest rivers that feed the Great Lakes and that were historically used by

Figure 2. Dam on Menominee River that prevents Lake Michigan lake sturgeon from migrating upstream. This dam is only 2.5 miles upstream from Lake Michigan and prevents access to more than 100 miles of previously utilized spawning and juvenile rearing habitat in the Menominee River. [Photo by D. Traynor, Michigan DNR.]

spawning lake sturgeon have dams that prevent use of historic spawning sites (Hay-Chmielewski and Whelan 1997). It should not come as a surprise that lake sturgeon populations have not rebounded from the dramatic declines that occurred in the late 1800s because the fish do not have access to the vitally important habitats they need to reproduce.

The detrimental impact of dam building and other habitat changes on lake sturgeon is perhaps best understood if we think of them as needing a large home range. Because lake sturgeon travel great distances in the open waters of the Great Lakes and up rivers for spawning, it should be rather obvious that the species needs large expanses of unimpeded habitat that could be considered their home range. However, few people seem to consider that large fish have these needs. All species of sturgeon travel over lake and river bottoms just as terrestrial organisms travel over field and forest. It is speculated that sturgeons move along habitual paths or corridors on lake bottoms and between lake systems such as exist in the Great

Lakes. Some fish species, and sturgeon specifically, migrate to avoid unfavorable conditions, to increase feeding opportunities, and to improve chances of finding a mate for spawning (Northcote 1978; Tsyplakov 1978; McKeown 1984). When migrating to spawn, lake sturgeon are known to return to historic spawning grounds and natal spawning beds (Auer 1999; Lyons and Kempinger 1992). Such homing behavior in fish is thought to help develop population-specific adaptations to the habitat (Leggett 1977) and can over time create distinct stocks. For instance, in the Volga River, Russia, eggs of three species of sturgeon differ in membrane strength and resistance to rupture from water pressure (a factor of depth and velocity of water movement) (Nikolsky 1963). The largest sturgeon of the three, the beluga sturgeon, *Huso huso*, spawned furthest upstream and had the most resistant egg membranes, which would be beneficial in the higher water velocities encountered in upstream reaches. Such adaptations may improve survival of eggs in environments of differing harshness.

There are 27 species of sturgeon known worldwide and these are categorized into four genera in the family Acipenseridae: *Huso* (2 sps.), *Acipenser* (19 sps.), *Scaphirhynchus* (3 sps.), and *Pseudoscaphirhynchus* (3 sps.) (Birstein 1993). Of all the sturgeon species known, only seven in the *Acipenser* and *Scaphirhynchus* genera are found in North America All sturgeons migrate in freshwater rivers to spawn, spawn in fast-flowing water, and are slow-growing, late-maturing fishes. All species take from 10 to 20 years to reach sexual maturity and most spawn intermittently. Three species of *Acipenser* are known to live their entire lives in freshwater. They include the Baikal sturgeon, *A. baeri baicalensis* (a subspecies of the Siberian sturgeon, *A. baeri*), which lives in Lake Baikal, Russia, and spawns in the tributaries of that system; the Yangtze sturgeon, *A. dabryanus*, from the lower Yangtze River, China; and *A. fulvescens*, the lake sturgeon known from three large drainage basins: the Great Lakes, Hudson Bay and the Mississippi River. The Yangtze sturgeon is considered very close to extinction (Doroshov and Binkowski 1985) particularly due to the construction of the Three Gorges dam (Wei et al. 1997). Of these three species, *A. baeri* and *A. fulvescens* occupy the coldest geographic ranges of any sturgeons, with *A. baeri* found throughout the major river systems of Siberia (Artyukhin 1995).

In organisms other than fish, a relationship between body size and range is usually positive (Gaston 1990). There is a positive relationship between body size and range for several fish species (McAllister et al. 1986), and a similar relationship between body size as mass (kg) and migration distance for birds, mammals, walking insects, and some fish (salmon) (Peters 1983). Mean age at first spawning as reflected in body size and distance of upriver migration for Atlantic salmon also shows a positive correlation (Schaffer and Elson 1975).

A similar relationship between body size and spawning migration distance is

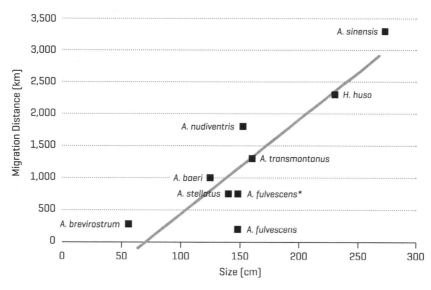

Figure 3. Relationship of average adult total length to river spawning migration distance for eight species of sturgeon. With lake sturgeon limited to 200 km, $r^2 = 0.785$; with lake sturgeon having maximum distance of 750 km (indicated by *), $r^2 = 0.874$ (adapted from Auer 1996b).

found in sturgeons (Auer 1996b). The body size as average total length of mature fish, and the maximum river spawning migration distance determined from values reported in the literature for historic or current spawning areas for sturgeons, show a positive relationship.

There is no record of possible total migration distance for lake sturgeon; because dams or other river barriers have long impacted many populations, such a distance for this species remains unknown. Similar-size sturgeons migrate farther in rivers than historical data show lake sturgeon migrating.

Comparing range and size data for all sturgeon species to those of an average adult lake sturgeon (about 145 cm TL, or 57 inches) it appears probable that this species could be capable of migrating distances of 1,000 to 1,800 km (620 miles). Current conditions to date show that lake sturgeon migrate shorter distances in rivers; relatively short segments of tributaries with natural barriers support lake sturgeon populations in the Great Lakes. In the Bad River, Wisconsin lake sturgeon migrate 32 km (20 miles) to spawn near natural falls and steep rapids (Shively and Kmiecik 1989). In the Grand River, Michigan lake sturgeon approach an impass-able set of rapids 64 km (40 miles) upstream, while in the Sturgeon River of upper

Michigan, lake sturgeon migrate 69 km (43 miles) to rapids about 1 km below a small hydroelectric facility (Auer 1996a).

These data indicate that the lake sturgeon, which live exclusively in freshwater, appear to combine river and lake distances in their spawning migrations. This is shown by migrations where lake sturgeon travel 228 km (141 miles) to spawn in the Lake Winnebago, Wisconsin, system, and they move 200 km (125 miles) within the Wolf River and can add 27 km (17 miles) as they cross Lake Winnebago (Lyons and Kempinger 1992). Auer (1999) found telemetry tracked postspawning lake sturgeon to leave the Sturgeon River, Michigan spawning habitat, 69 km (or 43 miles) upstream from the lake, and move distances of 265 km (165 miles) from the river mouth into Lake Superior. Adult lake sturgeon movements in the St. Lawrence River in Canada showed sturgeon moving distances of 138 to 225 km (86 to 140 miles) (Dumont et al. 1987; Fortin et al. 1993). Combining river and lake migration distances increases the closeness of the match of size to migration distance that is seen in other species.

Another reason open corridors for migration are so important for fishes is that spawning migrations are the time when hormones trigger the eggs and sperm to reach final development (McKeown 1984). Some species of female sturgeons, if prevented from reaching spawning areas because of barrier dams or manipulated flows, do not spawn and reabsorb eggs, or their spawned eggs show reduced survival (Artyukhin, Sukhoparova, and Fimukhina 1978; Veshchev and Novikova 1983, 1988). Scuba divers report sturgeon eggs, spawned below the Shawano Dam on the Wolf River in Wisconsin to be in masses as much as 15 cm (6 inches) thick (Kempinger 1988). Eggs in such masses are considerably more vulnerable to disease and predation. In the few remaining rivers where lake sturgeon can migrate onto natal spawning beds in natural rapids, with natural turbulent flows, their eggs adhere to the undersides of clean rock or fall into rock crevices, and thick masses of eggs are not observed. Years of research and monitoring have shown that lake sturgeon spawn in sections of rapids in the upper reaches of river systems and then return to feeding and wintering regions in lakes and river mouth reaches. Delta areas at the mouths of rivers are usually surrounded by rich organic sloughs or wetlands that are believed to provide food organisms and shelter from predators, necessary for the growth and survival of newly hatched and juvenile individuals.

Identified sites for rehabilitation and restocking of lake sturgeon should consider allocation to a minimum of 250–300 km (155 to 185 miles) of unrestricted movement, and migratory distances of 750–1,000 km (465 to 620 miles) should not be considered unusual for this species. For lake sturgeon, protecting open tributary corridors between river and lake environments or providing fish passage must be given priority, combined with safe outmigration pathways and areas for feeding and wintering of all life stages.

Other changes to Great Lakes habitat that may have affected lake sturgeon are associated with the logging era and industrial development. During the logging era, when sawmills were operating on many rivers, large quantities of sawdust were dumped directly into the rivers. This sawdust sank and in many places formed thick mats on the substrate of the river mouth and nearby Great Lakes bottomlands (Harkness and Dymond 1961). The mats of sawdust effectively eliminated large feeding areas vital to young sturgeon as they grew and developed in the lower reaches of rivers. Industrial pollution has also been widespread in the Great Lakes and contamination continues to plague some areas of the lakes. Lake sturgeon that inhabit areas of the lakes affected by industrial pollution have been shown to carry high body burdens of contaminants, including mercury, PCBs, dioxin, and others (Michigan DNRE unpublished data). The impacts of these contaminants on lake sturgeon are unknown, but reproductive impairment is possible.

Exotic species have also changed habitat characteristics in the Great Lakes. Most notable are the zebra and quagga mussels. Zebra mussels are native to the European and Russian Black, Azov, and Caspian seas, and quagga mussels are native to the Dnieper River and Ponto-Caspian Sea. Both of these species were probably introduced to the Great Lakes in the ballast water of ocean freighters. Zebra and quagga mussels are filter feeders and have widely colonized most of the available habitats in the lower lakes; to date Lake Superior remains relatively unaffected. The exotic mussels have dramatically increased water clarity by filtering out nutrients, phytoplankton, and zooplankton from the water column. The long-term impact of these changes on lake sturgeon is unknown, but the establishment of zebra and quagga mussels is believed to be responsible for the near-disappearance of *Diporeia* from the lower lakes (Nalepa, Fanslow, and Lang 2009). *Diporeia* is a small shrimplike animal and an important link in the Great Lakes food chain. Although lake sturgeon have been known to eat zebra and quagga mussels, there is nothing to suggest that the establishment of the exotic mussels and subsequent ecosystem changes are a good thing for lake sturgeon or any other fishes of the Great Lakes.

Foods and Feeding

As a general rule, as fish increase in size, the food items they consume also tend to increase in size. Therefore, as the largest fish found in the Great Lakes, lake sturgeon might be expected to eat the largest food items. However, lake sturgeon are an exception to the rule, and the foods they consume are among the smallest items available to them, such as insect larvae, small snails and clams, and crayfish.

One of the unique aspects of lake sturgeon form is the position of the mouth.

Most fish species found in the Great Lakes have a mouth positioned at or near the front of the head, in front of the eyes, and food is located by a combination of sight, sound, and smell. In contrast, the lake sturgeon mouth is located on the flat underside of the head directly beneath the eyes and is preceded by four sensory barbells.

The mouth of the lake sturgeon is also unique in that it is protrusible, that is, can be extended out like a vacuum hose.

Lake sturgeon locate food by swimming along the bottom of lakes and rivers and by dragging their sensitive barbels over the substrate. When the barbels encounter something edible, the mouth is quickly extended to the substrate and the prey is suctioned into the mouth. The prey can be detected by feel, taste, or the weak electric signals that are emitted by all living organisms. These weak electric signals are detected by sensory pits that are also located on the underside of the head (Boglione et al. 1999).

The type of food eaten is directly determined by the lake sturgeon's unique form and feeding habits, and their feeding habits change little as the fish grow throughout their lives. When newly hatched, they begin feeding on small zooplankton, and the smallest insect larvae available to them (Kempinger 1988). As the young lake sturgeon grow, the size of the insect larvae and zooplankton they consume also increases, but by the end of the first summer of life, when the young are 15–20 cm (6–8 inches) long, their diet is very similar to that of the adults that are 1.8 m (6 feet) long. As the young grow, they may add crayfish and clams to their diet when they become large enough to prey on them (Hay-Chmielewski 1987) but they also continue to eat small insect larvae and worms as they grow to very large size.

As lake sturgeon feed, they invariably ingest some of the substrate (sand, small stones, etc.) along with their prey. However, these fish are able to work prey around in their mouth and begin the process of mechanically breaking down the ingested material. At the same time they are able to sort out most of the undesired materials like sand and gravel and expel them prior to swallowing. Lake sturgeon do not have teeth, but they do have strong cartilaginous plates in the mouth cavity that can be ground together to reduce prey size. In addition, these fish have a muscular gizzard-like stomach that can further break up ingested prey. By grinding food in the mouth and in the gizzard, lake sturgeon are able to crush small clams, snails, crayfish, and other hard foods to make them easier to digest. After the prey is mechanically broken down in the mouth and stomach, digestion and absorption of nutrients continues in the spiral valve intestine.

The spiral valve intestine of lake sturgeon is similar to that found in sharks and is an adaptation that increases the surface area of the intestine and thus increases the absorption of nutrients as food is digested. In the spiral valve, a flap of tissue spirals down the length of the interior intestine wall much like a spiral staircase might follow the interior wall of a tower. This flap of tissue and the inner lining of the intestine absorb nutrients from digested food as the food passes down from the stomach.

Lake sturgeon apparently do not use sight at all to find food but instead rely entirely on their sensitive barbels and sensory receptors alone. Lake sturgeon are classified as opportunistic predators that will eat anything they encounter. Because their barbels have sensory capabilities to help locate food, the fish avoid areas where aquatic plants are growing. Rooted aquatic plants probably interfere with the food search, and so lake sturgeon cruise areas of bottom that allow unobstructed foraging. Although there have been only a few rigorous scientific investigations of lake sturgeon diet, they all indicate that lake sturgeon readily eat what is available and in approximate proportion to its availability in the environment. For example, in Black Lake, Michigan, adult lake sturgeon diet in 1986 included large numbers of *Hexagenia sp.* mayfly nymphs and *Chironomidae sp.* midge larvae (Hay-Chmielewski 1987). These two taxa represented 77.3 percent of the stomach contents of fish sampled. At the same time that lake sturgeon diet was being analyzed, samples of the substrate in Black Lake indicated that these two organisms were the most abundant food available. Although mayfly and midge larvae were the most abundant food items in the environment and in the diet, crayfish were also apparently important food for Black Lake sturgeon. Crayfish, although only 2.3 percent of the stomach contents by number, represented 66 percent of the stomach contents by weight. Similar diets have been found for other lake sturgeon populations (Choudhury, Bruch, and Dick 1996; Beamish, Noakes, and Rossiter 1998; Werner and Hayes 2004).

The ability of lake sturgeon to prey on a wide variety of food organisms has also been demonstrated by the few studies that have been conducted. As many as 42 different types of animals have been found in lake sturgeon stomachs from a single water body (Werner and Hayes 2004). Most of the prey eaten by lake sturgeon are invertebrates (insects, crustaceans, and worms), primarily mayfly and midge larvae, but the importance of particular prey varies by water body. In the St. Lawrence River, mayfly and midge larvae are eaten by lake sturgeon but the prey most frequently consumed is amphipods. The lake sturgeon diet may change to take advantage of prey that is seasonally abundant, as sturgeon have been known to feed on dead and dying gizzard shad during winter in Lake Winnebago, Wisconsin.

Lake sturgeon have also been shown to adapt their feeding patterns to changes in prey availability. Zebra and quagga mussels are exotic mussels now widely established in the Great Lakes and are causing dramatic changes in ecosystem structure and function (Nalepa, Fanslow, and Lang 2009). Lake sturgeon appear to be taking advantage of this new food resource and are eating these exotic mussels. Although it may seem beneficial for lake sturgeon to be feeding on the newly established mussels, the ecosystem disruption that has been caused by exotic species may be causing increased mortality in lake sturgeon and other species. Botulism has been confirmed as the cause of bird deaths in some places in the Great Lakes (Brand et al. 1988) and is also suspected of causing occasional die-offs of lake sturgeon in some parts of the Great Lakes (Perez-Fuentetaja, Lee, and Clapsadl undated) . The exact mechanism of botulism poisoning is not known, but is believed to be linked to the expansion of zebra and quagga mussels as well as the establishment of round and tubenose gobies, exotic fishes that have also become widespread in the Great Lakes in recent years.

Prior to scientific studies of the lake sturgeon diet there were a number of misconceptions about the role they played in the ecosystem. One of the early misconceptions was that lake sturgeon preyed on the eggs of more desirable fish species like walleye, trout, and salmon, and because of this, lake sturgeon were considered harmful to these species. However, in diet studies, the eggs of these and other game fishes have not been found in lake sturgeon stomachs. Furthermore, lake sturgeon are typically not found in the habitats that are used by walleye, trout, or salmon at the time these species are spawning, and thus it is unlikely that they would have the opportunity to prey on eggs of these species. Sturgeon are not considered a threat to such species and their reproductive success. Interestingly, the only known instances of lake sturgeon preying on relatively large quantities of fish eggs have occurred when they have consumed the eggs of their own species during spawning. In a few instances in the Sturgeon River, Michigan spawning males have been found to have lake sturgeon eggs in their feces when the fish have been captured for scientific study (Auer and Baker, personal observation). The same has been observed in the Black River, Michigan, and Wolf River, Wisconsin.

Lake sturgeon feed on the same prey that other fish such as white suckers and redhorse consume. Therefore, there is at least the opportunity for competition for prey. However, there have not been studies that document harm to either lake sturgeon or other species from competition for food resources. Because these species are native to the Great Lakes they have coexisted for thousands of years and are able to occupy similar habitats and utilize similar food resources.

During the course of feeding sturgeon do occasionally make mistakes. One of the most unusual items eaten by a sturgeon was reported from Germany and actually helped to solve a missing person's case. In an account detailed by Harkness and Dymond (1961) a German newspaper reported a large sturgeon was caught in 1927 that had in its stomach an Iron Cross, a German military honor medal, with a soldier's name on it. The soldier identified from the medal had been wounded and permanently disabled during fighting in World War I. The soldier had become despondent and had disappeared in 1920. Based on the discovery of the medal in the stomach of the sturgeon, the German police concluded that the soldier had committed suicide by jumping in the river. Other unusual items found in the stomachs of sturgeon included unopened sardine cans, a vanity case with a powder puff, cigarettes, coins, and fishhooks.

Conclusion

The precarious status of lake sturgeon in the Great Lakes is a concern of many natural resource management agencies around the lakes, and management agencies are actively pursuing lake sturgeon reintroduction and rehabilitation across the watershed. It is recognized that the continued low abundance of the species is a symptom of historic overharvest as well as the drastic habitat changes that have taken place. If lake sturgeon management efforts lead to increased abundance and the lake sturgeon again becomes a prominent species in the Great Lakes, it will be an indication that the Great Lakes ecosystem has regained some of its historic health and function.

REFERENCES

Artyukhin, E. N. 1995. On biogeography and relationships within the genus *Acipenser*. Sturgeon Quarterly 3(2): 6–7.

Artyukhin, E. N., A. D. Sukhoparova, and L. G. Fimukhina. 1978. The gonads of the sturgeon, *Acipenser guldenstadt*, in the zone below the dam of the Volograd water engineering system. Journal of Ichthyology 18:912–923.

Auer, Nancy A. 1996a. Response of spawning lake sturgeon to change in hydroelectric facility operation. Transactions of the American Fisheries Society 125(1): 66–77.

———. 1996b. Importance of habitat and migration to sturgeons with emphasis on lake sturgeon. Canadian Journal of Fisheries and Aquatic Sciences 553(suppl. 1): 152–160.

———. 1999. Population characteristics and movements of lake sturgeon in the Sturgeon River and Lake Superior. Journal of Great Lakes Research 25:282–293.

Auer, N. A., and E. A. Baker. 2002. Duration and drift of larval lake sturgeon in the Sturgeon River, Michigan. Journal of Applied Ichthyology 18:557–564.

Baldwin, N. S., R. W. Saalfeld, M. A. Ross, and J. J. Buettner. 1979. Commercial fish production in the Great Lakes 1867–1977. Technical Report No. 3, Great Lakes Fishery Commission.

Benson, A. C., T. M. Sutton, R. F. Elliott, and T. G. Meronek. 2005. Seasonal movement patterns and habitat preferences of age-0 lake sturgeon in the lower Peshtigo River, Wisconsin. Transactions of the American Fisheries Society 134:1400–1409.

Beamish, F. W. H., D. L. G. Noakes, and A. Rossiter. 1998. Feeding ecology of juvenile lake sturgeon, *Acipenser fulvescens*, in northern Ontario. Canadian Field-Naturalist 112:459–468.

Birstein, V. J. 1993. Sturgeons and paddlefishes: Threatened fishes in need of conservation.

Conservation Biology 7:773–787.

Boglione, C., P. Bronzi, E. Cataldi, S. Serra, F. Gagliardi, and S. Cataudella. 1999. Aspects of early development in the Adriatic sturgeon *Acipenser naccarii*. Journal of Applied Ichthyology 15:207–213.

Bott, K. 2006. Genetic analyses of dispersal, harvest mortality, and recruitment for remnant populations of lake sturgeon, *Acipenser fulvescens*, in open-water and riverine habitats of Lake Michigan. M.S. thesis, Michigan State University.

Brand, C. J., S. M. Schmitt, R. M. Duncan, and T. M. Cooley. 1988. An outbreak of type E botulism among common loons (*Gavia immer*) in Michigan's Upper Peninsula. Journal of Wildlife Diseases 24:471–476.

Chiotti, J. A., J. M. Holtgren, N. A. Auer, and S. A. Ogren. 2008. Lake sturgeon spawning habitat in the Manistee River, Michigan. North American Journal of Fisheries Management 28:1009–1019.

Choudhury, A., R. Bruch, and T. A. Dick. 1996. Helminths and food habits of lake sturgeon *Acipenser fulvescens* from the Lake Winnebago system, Wisconsin. American Midland Naturalist 135:274–282.

Diana, J. S., P. W. Webb, and T. Essington. 2003. Growth and appetite of juvenile lake sturgeon *Acipenser fulvescens*. Michigan Department of Natural Resources, Fisheries Research Report 2063.

Doroshov, S. I., and F. P. Binkowski. 1985. Epilogue: A perspective in sturgeon culture. *In*: North American sturgeons: Biology and aquaculture potential. F. P. Binkowski and S. I. Doroshov, eds. Dr. W. Junk Publishers.

Dumont, P., R. Fortin, G. Desjardins, and M. Bernard. 1987. Biology and exploitation of lake sturgeon in the Quebec waters of the Saint-Laurent River. *In*: Proceedings of a workshop on the lake sturgeon (*Acipenser fulvescens*). C. H. Olver, ed. Ontario Fisheries Technical Report Series No. 23.

Fortin, R., J. R. Mongeau, G. Desjardins, and P. Dumont. 1993. Movements and biological statistics of lake sturgeon (*Acipenser fulvescens*) populations from the St. Lawrence and Ottawa River system. Canadian Journal of Zoology 71:638–650.

Gaston, K. J. 1990. Patterns in the geographical ranges of species. Biological Reviews 65:105–129.

Harkness, W. J. K., and J. R. Dymond. 1961. The lake sturgeon. Ontario Department of Lands and Forests.

Hay-Chmielewski, E. M. 1987. Habitat preferences and movement patterns of the lake sturgeon (*Acipenser fulvescens*) in Black Lake Michigan. Michigan Department of Natural Resources, Fisheries Research Report 1949.

Hay-Chmielewski, L., and G. Whelan. 1997. Lake sturgeon rehabilitation strategy. Michigan Department of Natural Resources Fisheries Division Special Report No. 18.

Holey, M. E., E. A. Baker, T. F. Thuemler, and R. F. Elliott. 2000. Research and assessment needs to restore lake sturgeon in the Great Lakes. Great Lakes Fishery Trust, Workshop Results.

Holtgren, J. M., and N. A. Auer. 2004. Movement and habitat of juvenile lake sturgeon (*Acipenser fulvescens*) in the Sturgeon River / Portage Lake system, Michigan. Journal of Freshwater Ecology 19:419–432.

Houston, J. J. 1987. Status of the lake sturgeon, *Acipenser fulvescens*, in Canada. Canadian Field-Naturalist 101:171–185.

Kempinger, J. J. 1988. Spawning and early life history of lake sturgeon in the Lake Winnebago system, Wisconsin. *In*: Eleventh annual larval fish conference. R. D. Hoyt, ed. American Fisheries Society Symposium 5. American Fisheries Society.

LaHaye, M., A. Branchaud, M. Gendron, R. Verdon, and R. Fortin. 1992. Reproduction, early life history, and characteristics of spawning grounds of the lake sturgeon (*Acipenser fulvescens*) in Des Prairies and L'Assomption rivers, near Montreal, Quebec. Canadian Journal of Zoology 70:1681–1689.

Leggett, W. C. 1977. The ecology of fish migrations. Annual Review of Ecological Systems 8:285–308.

Lyons, J., and J. J. Kempinger. 1992. Movements of adult lake sturgeon in the Lake Winnebago system. Wisconsin Department of Natural Resources Research Publication RS-156- 92.

McAllister, D. E., S. P. Platania, F. W. Schueler, M. E. Baldwin, and D. S. Lee. 1986. Ichthyofaunal patterns on a geographic grid. *In*: The zoogeography of North American freshwater fishes. C. H. Hocutt and E. O. Wiley, eds. J. Wiley and Sons.

McKeown, B. A. 1984. Fish migration. Timber Press.

Nalepa, T. F., D. L. Fanslow, and G. A. Lang. 2009. Transformation of the offshore benthic community in Lake Michigan: Recent shift from the native amphipod *Diporeia* spp. to the invasive mussel *Dreissena rostriformin bugensis*. Freshwater Biology 54:466–479.

Nikolsky, G. V. 1963. The ecology of fishes. Academic Press.

Northcote, T. G. 1978. Migratory strategies and production of freshwater fishes. *In*: Ecology of freshwater fish production. S. Gerking, ed. John Wiley and Sons.

Perez-Fuentetaja, A., T. Lee, and M. Clapsadl. Undated. Botulism E in Lake Erie: Ecology and lower food web transfer. Final Report to U.S. Fish and Wildlife Service, Grant Award 26462.

Peters, R. H. 1983. The ecological implications of body size. Cambridge University Press.

Schaffer, W. M., and P. F. Elson. 1975. The adaptive significance of variations in life history among local populations of Atlantic salmon in North America. Ecology 56(3): 577–590.

Schram, S. T., J. Lindgren, and L. M. Evrard. 1999. Reintroduction of lake sturgeon in the St. Louis River, western Lake Superior. North American Journal of Fisheries Management 19:815–823.

Scott, W. B., and E. J. Crossman. 1973. Freshwater fishes of Canada. Fisheries Research Board of Canada Bulletin 184, Ottawa, Canada.

Shively, J. D., and N. Kmiecik. 1989. Inland fisheries enhancement activities within the ceded territory of Wisconsin during 1988. Administrative Report 89-1. Great Lakes Indian Fish and Wildlife Commission.

Smith, K. M., and D. K. King. 2005. Dynamics and extent of larval lake sturgeon *Acipenser fulvescens* drift in the Upper Black River, Michigan. Journal of Applied Ichthyology 21:161–168.

Threader, R. W., R. J. Pope, and P. R. H. Schaap. 1998. Development of a habitat suitability index model for lake sturgeon (*Acipenser fulvescens*). Ontario Hydro Report Number H-07015.01-0012.

Thuemler, T. F. 1985. The lake sturgeon, *Acipenser fulvescens*, in the Menominee River, Wisconsin-Michigan. Environmental Biology of Fishes 14:73–78.

Tsyplakov, E. P. 1978. Migrations and distribution of the sterlet, *Acipenser ruthenus*, in Kuybyshev reservoir. Journal of Ichthyology 18:905–912.

Veshchev, P. V., and A. S. Novikova. 1983. Reproduction of the stellate sturgeon, *Acipenser stellatus* (Acipenseridae), under regulated flow conditions in the Volga River. Journal of Ichthyology 23(1): 42–50.

———. 1988. Reproduction of sevryuga, *Acipenser stellatus*, in the lower Volga. Journal of Ichthyology 28(1): 39–47.

Wehrly, K. W. 1995. The effect of water temperature on the growth of juvenile lake sturgeon *Acipenser fulvescens*. Michigan Department of Natural Resources, Fisheries Research Report 2004.

Wei, Q., F. Ke, J. Zhang, P. Zhuang, J. Luo, R. Zhou, and W. Yang. 1997. Biology, fisheries, and conservation of sturgeon and paddlefish in China. Environmental Biology of Fishes 48:241–255.

Werner, R. G., and J. Hayes. 2004. Contributing factors in habitat selection by lake sturgeon (*Acipenser fulvescens*). Final report submitted to Environmental Protection Agency Great Lakes National Program Office.

Plate 1. Scutes on juvenile sturgeon are razor sharp and run along the back, sides, and bottom edge on each side of the fish. [N. Auer.]

Plate 2a. The mouth of a sturgeon has no teeth. [N. Auer.]

Plate 2b. Sturgeon locate food using the four sensitive barbels located just in front of the mouth as they swim over the lake bottom. [N. Auer.]

Plate 3. Eggs of lake sturgeon, which adhere to clean rocks in rapidly flowing, oxygen-rich water. [N. Auer.]

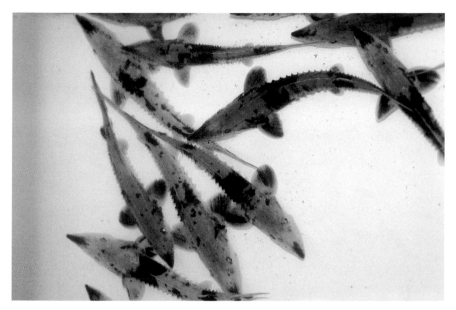

Plate 4. Young juveniles showing dark saddle markings on back and rich copper color. [N. Auer.]

Plate 5. Lake sturgeon spawning rapids in the Sturgeon River, Michigan. [Photo by K. Koval, Michigan DNR.]

Plate 6. Juvenile lake sturgeon shown with characteristic mottling that helps the fish blend in against sandy and mixed sand/gravel substrates. [Photo by E. Baker, Michigan DNR.]

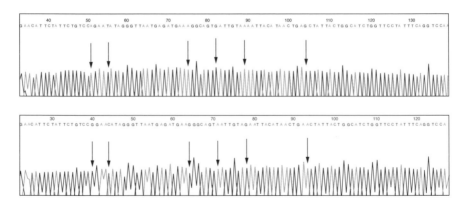

Plate 7. Mitochondrial DNA sequences from two lake sturgeon individuals. The individual on the top is from the Detroit River and the individual on the bottom is from the Lower Niagara River. Arrows indicate polymorphisms, or places in the sequence where the two individuals differ.

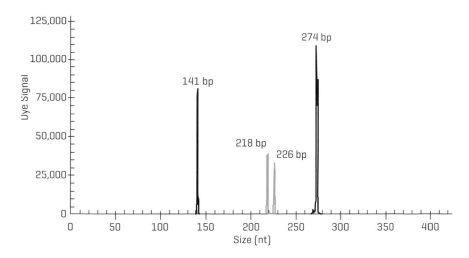

Plate 8. Microsatellite data from a single lake sturgeon individual from the Namakan River, Ontario.

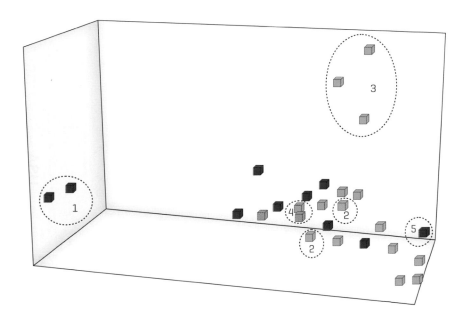

Plate 9. Stock structure of lake sturgeon in the Great Lakes basin. Each cube represents a spawning population. The different colors represent the different lake basins: red, Lake Superior; green, Lake Michigan; blue, Lake Huron; light purple, Lake Ontario; dark purple, St. Lawrence River system; orange, Lake Champlain. Circled and numbered regions are described in more detail in the text.

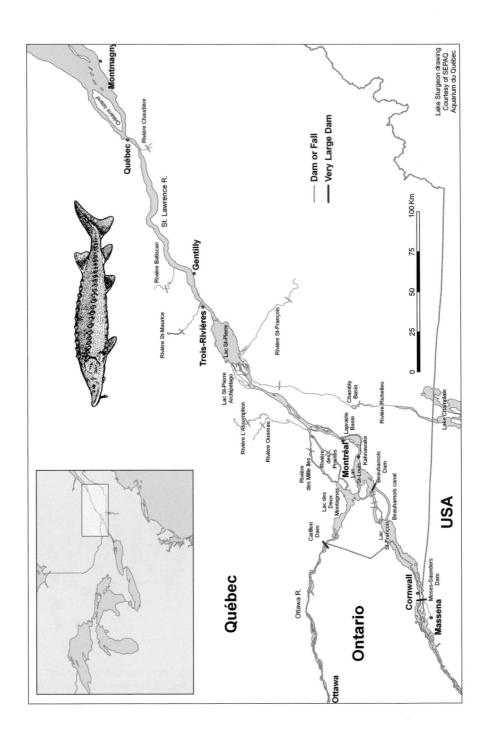

Québec

Ontario

Ottawa

Ottawa R.

Carillon Dam

Lac des Deux Montagnes

Rivière des Mille Îles

Rivière des Prairies

Montréal

Lac St-Louis

Kahnawake

Beauharnois Dam

Beauharnois canal

Lac St-François

Cornwall

Massena

Moses-Saunders Dam

USA

Lake Champlain

Laprairie Basin

Chambly Basin

Rivière Richelieu

Rivière St-François

Rivière L'Assomption

Rivière Ouareau

Lac St-Pierre Archipelago

Lac St-Pierre

Trois-Rivières

Rivière St-Maurice

Gentilly

Rivière Batiscan

St. Lawrence R.

Québec

Rivière Chaudière

Orléans Island

Montmagny

— Dam or Fall

— Very Large Dam

0   25   50   75   100 Km

Lake Sturgeon drawing
Courtesy of SEPAQ
Aquarium du Québec

Plate 10 [*opposite*]. Major habitats of the lake sturgeon populations in the Quebec part of the St. Lawrence River system. Water discharge averages 7,500 m³/s in front of Cornwall, 8,000 m³/s in front of Montreal, and 12,600 m³/s in front of Quebec city. The average discharge of the Ottawa River at Carillon dam, at the head of Lac des Deux Montagnes [158 km²] is 2,000 m³/s. Water depth is generally low [< 3 m] except in the navigation channel [minimum 11.2 m] and in a few deeps in Lac Saint-François [26 m], Lac Saint-Louis [27 m], and Lac des Deux-Montagnes [47 m]. Lake sturgeon commercial fishing is allowed in Lac Saint-Louis [total area, 148 km²], in Laprairie Basin [43 km²], in a short reach downstream Montreal, in Lac St-Pierre [381 km²], and in the upper estuary. Elsewhere, only sport fishing is authorized, except in Lac Saint-François [272 km²], where both fishing activities have been prohibited since 1987.

Plate 11a and 11b. Current Odanak Abenakis Council armorial bearings [a], also represented as a vegetal sculpture during the Montreal's Mosaicultures in 2000 and 2001 [b].

Plate 12. Streamside rearing trailer located on Manistee River, Michigan.

Plate 13. The newly-released fish blend into the river background due to their cryptic color pattern.

AMY WELSH

Recognizing the Genetic Population Structure of Lake Sturgeon Stocks

MIGRATION PATTERNS AND THE PRESENCE OF CORRIDORS PLAY A LARGE ROLE IN defining different stocks of fish. Groups of fish may be frequently migrating between different locations and may even be reproducing with the fish they encounter at the new location. On the other hand, fish may move around quite a bit when not spawning, but when it comes time to reproduce, they go to the same spot each time. It can often be hard to detect migration (with resulting reproduction at the new location) through field techniques. That's where theories of population genetics can help.

If populations are not connected through migration, then genetic differences can accumulate between them. This can occur in two primary ways. First, genetic differences may result from chance. Picture a bowl of M&M candies. The colors represent different variants of the same gene (called alleles). If you picked 20 M&Ms from the bowl to represent Population 1, calculated the allele frequencies, and then repeated the process for Population 2, you'd probably find that the allele frequencies would be different for the two populations. And this was due purely to chance. Now if your two M&M populations never mated with each other, then those allele frequencies would remain different throughout the generations. However, if they did mate with each other, eventually the allele frequencies would be the same in the two populations.

Another way differences between populations can result is through natural selection. Certain alleles may produce a trait in a fish that is beneficial in terms of survival or reproduction. For example, in many bird species, selection will favor alleles that result in bright colors because that trait gives the individual reproductive advantages. If different traits are favored in different environments, then natural selection may result in allele frequency differences between populations. Migration would have little effect on these traits because migrants with a trait not suited for its new environment would feel the wrath of natural selection.

So why do we care whether or not sturgeon move between spawning locations for reproduction? The main reason is genetic diversity. It is a good thing to have high genetic diversity. With high genetic diversity, natural selection has many genes at its disposal to give the species an evolutionary advantage. Environments change and it is critical species have the ability to adapt. If there are only a few alleles for a certain gene in a population, and none of them are doing the trick, evolution has to wait for a mutation to occur, and that can take a long time. Instead, if a population has lots of alleles, it is more likely that an allele already exists that is best suited for the current environment.

Inbreeding (the mating of close relatives) can reduce the genetic diversity in a population. If populations are small and isolated, then they may have little choice but to mate with a relative. Royal families in the past who condoned this practice in order to maintain "royal blood" felt the consequences through numerous afflictions. Inbreeding increases the chances that an individual is going to have two copies of the same allele for a particular gene (i.e., homozygosity). For many alleles that result in detrimental traits, two copies of the bad allele are necessary in order to exhibit the trait. A common example is sickle cell anemia. Those with the disease get a copy of the bad allele from both their mother and their father.

Inbreeding can have devastating consequences in many animal and plant species. Fish are no exception. In a study of the threespine stickleback, both survival and reproductive success were significantly impacted by one generation of brother-sister matings (Frommen et al. 2008). Inbred families of Atlantic salmon also had decreased survival compared to non-inbred families (Ryman 1970). Reduced survival and growth have also been observed in hatchery-reared Pacific salmon (Kincaid 1983).

Genetic diversity may exist in a species as a whole through the maintenance of different stocks. If a fish species consists of several stocks representing different lineages, then the genetic differences between stocks contribute to the diversity of the species as a whole. By assessing the genetic connectivity of stocks, managers can maintain linkages between connected stocks. Stock differentiation may be maintained in the species through natal homing, where individuals return for spawning to the stream where they were born. Management actions should try to maintain the natural stock structure. There may be a reason that stock structure has evolved, and it's probably better if human actions don't interfere.

One of the known consequences of disregard for natural stock structure is outbreeding depression. This can occur when individuals from different species or even genetically distinct populations mate. The resulting offspring may be less fit than either one of the parents. The best-known example is the mule. Mules result from the mating of a donkey and a horse, and they are sterile. Outbreeding depression has also been observed in fish. In rainbow trout, offspring resulting from matings between hatchery-produced and natural stocks had lower survival during the winter than individual natural stocks (Miller, Close, and Kapuscinski 2004). Changes in embryo development time and survival were observed in progeny from the matings of three geographically separate stocks of southeast Alaska coho salmon (Granath et al. 2004). In pink salmon, decreased survival of crosses between genetically distinct populations also provide evidence for outbreeding depression (Gilk et al. 2004). Outbreeding between largemouth bass populations with moderate levels of genetic differentiation resulted in increased infectious disease susceptibility (Goldberg et al. 2005).

In between the extremes of outbreeding and inbreeding depression is the middle ground that should be the goal of fishery managers. Some level of outbreeding is necessary to maintain the genetic diversity of populations and to avoid the effects of inbreeding. Hybrid vigor, or heterosis, occurs when the hybrid has greater fitness than either one of the parent strains. Examples of heterosis include increased growth in hatchery-raised silver perch (Guy, Jerry, and Rowland 2009) and increased body weight in hybrids from different tilapia strains (Maluwa and Gjerde 2006). Mating of individuals from genetically different populations can also increase genetic diversity by increasing heterozygosity (i.e., having two different alleles) and by introducing new alleles into the population. Increased heterozygosity can be especially important at the major histocompatibility complex (MHC), for example. The MHC plays an important role in immune defense in vertebrates. There is strong selection for heterozygosity at the MHC because heterozygotes can respond to a larger number of pathogens. The large number of alleles resulting from outbreeding can also provide more genetic material with which selection can work. Instead of waiting around for mutation to generate new genetic combinations, outbreeding can create those potentially favorable combinations much more quickly.

Genetic Techniques

Scientists can take various approaches to assess the genetic diversity in populations. The first step involves deciding which part of the genome should be analyzed to adequately answer the research question. For genetic questions about stock structure, two genetic markers are generally used: mitochondrial DNA and nuclear DNA through the use of microsatellites.

MITOCHONDRIAL DNA

Cells contain many mitochondria, which are structures within the cell that are essentially the powerhouse for the cell. Within each mitochondrion, there are several copies of circular DNA. This mitochondrial DNA (mtDNA) has several unique characteristics that make it different from the DNA in the nucleus of the cell. First, mtDNA is maternally inherited, which means that, barring any mutations, one has the exact same mtDNA as one's mother. The father generally does not contribute any mtDNA to his offspring. Second, each mitochondrion has several copies of mtDNA and each cell has several mitochondria. Therefore, each cell has many mtDNA copies. On the other hand, DNA housed in the nucleus of the cell is only present in a single copy and there is only one nucleus per cell. This makes mtDNA particularly useful for studying ancient samples where DNA degradation may have occurred. Third, nuclear DNA experiences recombination, where portions of the chromosomes inherited from the mother and father are swapped. This does not occur in mtDNA; therefore, all genes on the mtDNA are linked together. Many genes on the mtDNA experience strong natural selection and, because of the lack of recombination, all other genes on the mtDNA may also appear as though they are experiencing natural selection.

To study mtDNA, researchers generally compare the sequences of DNA. DNA consists of four different components called bases, coded by the letters A, G, C, and T. Individuals can vary in the order or sequence of the bases, and that's how genetic diversity is measured using mtDNA. Plate 7 depicts sequence data from a portion of the mtDNA from two lake sturgeon. The different colors represent the different bases. The arrows point to locations where the two individuals differ in their mtDNA sequence. These differences are called polymorphisms.

NUCLEAR DNA AND MICROSATELLITES

DNA in the nucleus of the cell is arranged differently than mtDNA. Instead of being a circular piece of DNA, it is arranged linearly into chromosomes. Humans and most other animal species have two copies of each chromosome: one copy comes from the mother and one copy from the father. Sturgeon species, however, are rather different because many of the species are polyploid. Instead of having two chromosomal copies, they may have four (e.g., lake sturgeon) or even eight (e.g., shortnose sturgeon) chromosomal copies (Ludwig et al. 2001). This phenomenon is also observed in many plant species, a few amphibians, and a few other fish species. Interestingly, it appears that sturgeon are evolving to having only two copies of each gene. In lake sturgeon, while the majority of genes have four copies, approximately 25 percent of the genes have been reduced to two copies (Welsh, Blumberg, and May 2003).

Microsatellites are short repeated sequences of DNA. The repeats consist of two to four bases. Here is an example of a DNA sequence containing microsatellites:

ATCGATGCGATA GATA GATA GATA GATACCCTCA

Note how the sequence GATA is repeated several times. GATA is an example of a microsatellite, and this sequence contains five repeats of the microsatellite. However, other individuals may have only three or four repeats, while some individuals may have six or seven repeats. When comparing the same location of the nuclear DNA, the different number of repeats results in different variants or alleles that can be used to measure genetic diversity.

Researchers will examine several positions or loci throughout the nuclear DNA-containing microsatellites and, instead of directly sequencing the DNA, will use a more indirect method of detecting microsatellite variation. If you were looking at a fragment of DNA containing microsatellites, the length of the fragment would depend on the number of microsatellite repeats it contained. Fragments with more repeats would be longer than fragments with fewer repeats. Plate 8 shows microsatellite data from a lake sturgeon from the Namakan River in Ontario. There are data from three different loci represented in the plate, and these are denoted by the different colors. The peaks represent the alleles and are separated out by size, with the peaks to the far left containing fewer bases than the peaks to the far right. Size is measured in the number of bases (bp). Each locus can have one or two peaks. If the locus has one peak (e.g., the black and blue loci), that individual received the same allele from both its mother and father and is called a homozygote. If the locus has two peaks (e.g., the green locus), that individual received a different allele from its mother and father and is called a heterozygote. The number of alleles in a population and the proportion of loci that are heterozygous (called heterozygosity) are measures of genetic diversity.

SUMMARY OF GENETIC TECHNIQUES

Both mtDNA and microsatellites have their advantages in helping determine diversity, depending on the research question being addressed. Mitochondrial DNA has a slower mutation rate than microsatellites, so it is useful for studying differences between species or for understanding the impact of historical influences on stock structure. Mitochondrial DNA is limited because it may be under the influence of selection, which may confound any conclusions about movement patterns. Because it is maternally inherited, any movement patterns inferred would only be relevant to movements of the females in the population. Microsatellites, with their relatively high mutation rate, are optimal for detecting population differences within a species or for detecting stock structure influenced by evolutionarily recent events. These repeated sequences of DNA may be "junk DNA" that don't code for any proteins. Therefore, they may be less influenced by selection, and differences between stocks are more likely due to a lack of gene flow.

North American Stocks of Lake Sturgeon

The above genetic techniques have been used to evaluate lake sturgeon stock structure throughout the Great Lakes, the Hudson Bay / James River, and the Mississippi River watersheds. Many laboratories have been working on different regions of the lake sturgeon's range and, through extensive coordination, data have been compiled to obtain a rangewide picture of stock structure.

RANGEWIDE STOCK STRUCTURE

An early study examined lake sturgeon genetic diversity throughout all three watersheds, comparing sturgeon from the northern parts of their range to the southern parts using mtDNA (Ferguson and Duckworth 1997). In the portion of mtDNA examined, only two different sequences were identified. Both sequences were found in lake sturgeon of the Hudson Bay / James Bay watershed. However, only a single sequence predominated in lake sturgeon from the Great Lakes and the Mississippi watershed. It became clear that the sturgeon from Hudson Bay / James Bay represented a different stock from the rest of its range. Because mtDNA is especially useful in determining historical influences on stock structure, differences between the watersheds could possibly be attributed to the use of different refugia during glaciation. The study also indicated that lake sturgeon had relatively low genetic diversity, as indicated by the low number of different sequences. The Great Lakes seemed particularly affected by low genetic diversity with the presence of only a single sequence.

Subsequent studies used microsatellites to assess genetic differences between lake sturgeon from two populations in the Hudson Bay / James Bay and several populations in the Great Lakes watersheds (McQuown et al. 2003; Welsh et al. 2008). Using this genetic technique, high genetic diversity (high heterozygosity and a high number of alleles) was observed in both watersheds, and these studies also confirmed the genetic differences observed between the two watersheds. A common measure of genetic differentiation between populations is F_{ST}. F_{ST} values can range from 0 to 1, with 0 indicating the populations are the same and 1 indicating they are completely different. However, values as high as 1 are rarely observed. In fact, values of 0.15–0.25 indicate strong genetic differentiation between populations (Wright 1978). The average F_{ST} value between populations from the Hudson Bay / James Bay and the Great Lakes is 0.18 (Welsh et al. 2008). Therefore, these results confirm that sturgeon in the two watersheds are very different.

A recent study has compared lake sturgeon from all three watersheds using both microsatellies and mtDNA (Drauch et al. 2008). Sampling focused on spawning locations located in the midwestern United States. Again, the greatest genetic differences were observed between watersheds. Population differences were highest between the Hudson Bay / James Bay watershed and the other watersheds. Samples from the Mississippi and Missouri rivers, however, were not much different from the Great Lakes. The reason for this similarity is that the lake sturgeon in those two rivers have mainly been reintroduced into the system and the source for the reintroduction was lake sturgeon from Lake Winnebago (below Green Bay in Lake Michigan). Some remnant populations do remain in the Mississippi watershed. The lake sturgeon population in the White River of the Ohio River drainage does not appear to have a heavy influx of migrants from the reintroduced fish. Also, the Chippewa River, a tributary of the upper Mississippi River, seems free of Lake Winnebago migrants, likely because of an impoundment that may limit immigration.

In summary, the three different watersheds (Hudson Bay / James Bay, Great Lakes, and Mississippi River) clearly represent three different stocks, with the highest genetic differentiation observed between the Hudson Bay / James Bay and the other two watersheds. The level of differentiation of the Hudson Bay / James Bay lake sturgeon indicates these fish are likely headed down a different evolutionary path and may eventually be designated as subspecies. Lake sturgeon in the Mississippi River watershed are also distinct from the other two watersheds. However, through reintroductions from the Great Lakes system, the natural stock structure may be altered.

GREAT LAKES STOCK STRUCTURE

Within the Great Lakes watershed itself, significant stock structure exists. Contrary to early studies indicating low genetic diversity in Great Lakes lake sturgeon populations, new research shows a high level of diversity. The level of genetic differentiation varies between lake basins, but throughout the Great Lakes, most lake sturgeon spawning populations are genetically distinct from each other (DeHaan et al. 2006; Welsh et al. 2008). Therefore, most lake sturgeon spawning populations represent individual stocks. The most likely reason for the genetic distinctions observed between spawning populations is natal homing. Lake sturgeon are most likely returning to the stream where they were born in order to spawn. The low migration (with resulting reproduction) between spawning populations results in genetic differentiation between locations. Natural selection is likely not the reason for the observed differences because microsatellites are considered to be neutral genetic markers (i.e., noncoding DNA that is not under the influence of selection).

Plate 9 summarizes the lake sturgeon stock structure in the Great Lakes watershed. Each cube represents a sampled spawning location, with colors corresponding to their respective lake basin. Those populations closest together in the plate are most genetically similar. There are some interesting points to note about the stock structure. The numbered points below correspond to the numbered circles in the plate.

1. These two populations are from the Bad and White rivers along the south shore of Lake Superior. They are different from the rest of Lake Superior and the rest of the Great Lakes. In fact, their level of genetic differentiation is almost as high as the genetic differentiation between the Great Lakes and Hudson Bay / James Bay. In general, much higher genetic differentiation was observed between populations within Lake Superior than observed elsewhere in the Great Lakes (Welsh et al. 2008). Also, Lake Superior is highly differentiated from the rest of the Great Lakes.

There are several possible reasons for the high genetic distinction of Lake Superior. It could be due to the glacial history of the region. The Great Lakes were covered by glaciers 8,000–10,000 years ago. During that time, fish species had to move to several different refugia. Once the glaciers receded, fish could recolonize the Great Lakes. However, during that time in separate refugia, genetic differences may have accumulated. Artifacts from glacial refugia could still be present in current genetic data since 10,000 years ago is not that long from an evolutionary perspective.

If the Lake Superior lake sturgeon inhabited a glacial refugia different from the rest of the Great Lakes sturgeon, this may help to explain some of the genetic differences observed.

The genetic distinction of populations within Lake Superior is likely due to the geography of the lake. Lake Superior is the largest in surface area and deepest Great Lake and in several areas, depths of several hundred feet are found close to shore. Lake sturgeon prefer relatively shallow depths. Therefore, movement along the shoreline may be limited, preventing the dispersal between spawning locations.

2. The two green cubes circled in plate 9 represent the Manistee and Muskegon rivers, whose mouths are on the eastern shore of Lake Michigan. These two lake sturgeon populations are more closely related to spawning populations from Lake Huron (blue cubes) than they are to populations in Green Bay (green cubes), on the western side of Lake Michigan (DeHaan et al. 2006). This is an important example that illustrates that sturgeon don't adhere to human-created boundaries. Most management in the Great Lakes is conducted through coordination within each lake basin. For example, the Great Lakes Fishery Commission, which helps coordinate activities throughout the Great Lakes, is structured through lake committees. However, in some cases, this may not be applicable. In the case of the lake sturgeon, coordination would be necessary between Lake Michigan and Lake Huron to accommodate the stock structure of lake sturgeon.

3. The circled blue cubes represent a population from the eastern side of Lake Nipissing, a population from the western side of Lake Nipissing, and the Spanish River (Welsh et al. 2008). These spawning locations lie off Georgian Bay in Lake Huron and group apart from other Lake Huron populations, which include the Mississauga River, the Detroit River, and the St. Clair River. The Mississauga River also feeds Georgian Bay, and other analyses have a difficult time grouping it with any other spawning populations. The Detroit / St. Clair system connects Lakes Huron and Erie.

4. Two spawning populations from different lake basins are very similar to each other: the St. Clair River and the Lower Niagara River, which is below Niagara Falls and flows into Lake Ontario. The Detroit / St. Clair lake sturgeon are genetically indistinguishable from the Lower Niagara sturgeon (Welsh et al. 2008), and this is surprising. There's quite a bit of distance separating the two locations (in fact, the whole of Lake Erie) and a substantial natural barrier to gene flow (i.e., Niagara Falls). Surprisingly, fish fare relatively well going down Niagara Falls. The Welland Canal also provides a way around the falls, but there is little evidence of fish regularly using the canal as a dispersal route (Daniels 2001).

History at the Lower Niagara River may provide insight into the similarities between these spawning locations. The Lower Niagara population supported both commercial and recreational fisheries until the 1940s, when the population drastically declined and fisheries collapsed (Aug 1992). Between 1980 and 1994, only two lake sturgeon were known to be captured in the Lower Niagara River (Carlson 1995). A more recent study confirmed low lake sturgeon abundance in the river (Hughes, Lowie, and Haynes 2005). Juvenile age classes dominated the population, with individuals ranging from ages of 1 to 23 years and the majority below 10 years of age (Hughes, Lowie, and Haynes 2005). These age data provide evidence for a new or recovering population. Therefore, the lack of genetic differentiation between the Detroit / St. Clair River and the Lower Niagara River could be due to a recent recolonization of the Lower Niagara by sturgeon from Lake Huron or Lake Erie.

5. The dark purple cube circled in plate 9 is the Grasse River, a tributary of the St. Lawrence River. Although the point is not as clear in this plate, this spawning population is highly distinct from the rest of the Great Lakes populations. It also has low heterozygosity and a low number of alleles relative to other Great Lakes populations. The reason for this may be its isolation. A dam that had been in place on the Grasse River may have prevented migrants from entering. That leaves a small isolated population above the dam. Small isolated populations start to lose alleles through chance, and they have to start mating with their relatives because there are few alternatives. The high differentiation observed at the Grasse River is likely due to chance effects from isolation, as evidenced by the population's low level of genetic diversity.

Management Implications

MOVEMENT PATTERNS DURING NONSPAWNING PERIODS

Once the stock structure of a fish has been defined, that information can be used to do some detective work. Lake sturgeon spend most of their lives doing everything but spawning. We have genetic data about their spawning times, but don't know much about what they're doing or where they're going when not spawning. This can have important implications for management, because activities in one jurisdiction can have an impact on spawning populations from a different jurisdiction. For example, in some states and provinces, fishing of lake sturgeon is still permitted in certain waters. Fishing targets nonspawning sturgeon, but we don't know which stocks are most affected by the harvest. Therefore, management actions focused on the spawning population may be less effective if those same fish are being harvested somewhere else. For example, a lake sturgeon recreational fishery remains on the Menominee River, and the genetic data on stock structure were used to identify which stocks were being harvested. Although 82 percent of the harvested sturgeon were from the Menominee, 18 percent were from the Oconto/Peshtigo rivers (Bott et al. 2009).

DAMS

Hydroelectric facilities can affect various lake sturgeon life stages, including the spawning period. Understanding stock structure can help managers understand which groups are still connected genetically and, one hopes, maintain those groups. Dams can obviously break those connections. As seen with Niagara Falls, it is usually not hard for sturgeon to get down a dam. Getting back up is the problem. As a result, genes can move out of the population, but no new genes come into these populations. There usually is a population of sturgeon that will remain above the dam, creating a pseudo-"river resident" population. Examples of this have occurred on the Menominee River and the Grasse River. As mentioned, the population on the Grasse River has lower heterozygosity and fewer alleles than other Great Lakes populations. This is likely because the dam created a small isolated population above the dam, where individuals may leave but cannot return. Now that the dam has been removed, migrants can come to the Grasse to spawn. This could increase the genetic diversity of the population. It doesn't take much. In general, one migrant per generation can do the trick. Permitting migration could also rescue inbred populations.

STOCKING

Many lake sturgeon spawning populations throughout the basin have gone extinct, and some management agencies have been interested in taking a more active approach to restoring lake sturgeon to their waters. One approach that can speed up reintroduction to historic spawning locations is stocking. With stocking, agencies can reach management goals more quickly than if they wait for natural recolonization to occur. For example, in the state of New York, lake sturgeon are listed as threatened. In order to be delisted, the state needs to have eight self-sustaining populations of lake sturgeon. Therefore, the state has implemented a fairly extensive stocking program. However, in light of the stock structure observed in lake sturgeon, great care needs to be exercised before fish are reintroduced widely.

Most lake sturgeon spawning populations represent a unique stock. As a result, if different stocks are mixed, outbreeding depression could occur. Isn't genetic diversity good? Sometimes it's a good idea, but there is a fine line between inbreeding depression (poor survival and reproduction due to mating with close relatives) and outbreeding depression (poor survival and reproduction due to mating between very different individuals). The case of the Florida panther presents an example of benefits from mixing different gene pools. This species lives in the Everglades and declined to a small population size in the wild (60–70 individuals) (Frankham, Ballou, and Briscoe 2002). The population was clearly showing signs of inbreeding depression (e.g., heart defects, poor sperm quality), so biologists decided to introduce females from a Texas subspecies. This resulted in improved fitness of the panthers. When there is no evidence of inbreeding, it is best to stick with similar stocks for mating. In the case of lake sturgeon, there is no evidence for inbreeding and small populations do not appear to have any less genetic diversity than large populations (DeHaan et al. 2006). Therefore, the best management strategy is to conserve the natural stock structure.

If inbreeding becomes a problem for the small lake sturgeon populations that remain, then stocking is a management strategy that increases the genetic diversity of a population. Individuals from other nearby populations could be introduced, or hatchery matings of sturgeon could occur between individuals that are known to be unrelated. Care needs to be exercised when stocking because sometimes stocking can actually increase the level of inbreeding. If only a few individuals are mated and large numbers of related offspring are released, then the level of relatedness in the population could actually increase.

Because of strong differences between watersheds, selection of source populations at a minimum should remain within the same watershed. Genetic differences between certain locations in the Mississippi River watershed and the Great Lakes have already been reduced through stocking with individuals outside of the native watershed. When selecting a stock to use as a source for a stocking program, the goal should be to select a stock most genetically similar to sturgeon in the vicinity of the stocking location to avoid the risk of outbreeding depression in the small natural populations that remain. Lake sturgeon were stocked in the St. Louis River of Lake Superior in the 1980s, and initially sturgeon from the Wolf River (Green Bay, Lake Michigan) were used. Later, it was determined that it would probably be better to use a source within Lake Superior, and the source was then switched to the Sturgeon River on the south shore of Lake Superior (Schram, Lindgren, and Evrard 1999).

Conclusion

Population genetics can provide an important tool in understanding stock structure and complementing other ecological studies on movement patterns. Understanding stock structure is also about understanding genetic diversity, the fuel that keeps evolution moving along. Recent research indicate that lake sturgeon have substantial stock structure throughout their range. The Hudson Bay / James Bay, Great Lakes, and Mississippi River watersheds are highly differentiated. Within the Great Lakes itself, most spawning populations represent a unique stock. As humans continue to modify their environment and as management agencies seek to remedy environmental problems, it is important to remember the genetic differences that exist among lake sturgeon stocks and make every effort to conserve this important source of genetic diversity.

REFERENCES

Aug, L. 1992. Beyond the falls: A modern history of the lower Niagara River. Niagara Books.

Bott, K., G.W. Kornely, M.C. Donofrio, R.F. Elliott, and K.T. Scribner. 2009. Mixed-stock analysis of lake sturgeon in the Menominee River sport harvest and adjoining waters of Lake Michigan. North American Journal of Fisheries Management 29:1636-1643.

Carlson, D. M. 1995. Lake sturgeon waters and fisheries in New York State. Journal of Great Lakes Research 21:35–41.

Daniels, R. A. 2001. Untested assumptions: The role of canals in the dispersal of sea lamprey, alewife, and other fishes in the eastern United States. Environmental Biology of Fishes 60:309–329.

DeHaan, P. W., S. T. Libants, R. F. Elliott, and K. T. Scribner. 2006. Genetic population structure of remnant lake sturgeon populations in the upper Great Lakes basin. Transactions of the American Fisheries Society 135:1478–1492.

Drauch, A. M., B. E. Fisher, E. K. Latch, J. A. Fike, and O. E. Rhodes Jr. 2008. Evaluation of a remnant lake sturgeon population's utility as a source of reintroductions in the Ohio River system. Conservation Genetics 9:1195-1209.

Ferguson, M. M., and G. A. Duckworth. 1997. The status and distribution of lake sturgeon, *Acipenser fulvescens*, in the Canadian provinces of Manitoba, Ontario and Quebec: A genetic perspective. Environmental Biology of Fishes 48:299–309.

Frankham, R., J. D. Ballou, and D. A. Briscoe. 2002. Introduction to conservation genetics. Cambridge University Press.

Frommen, J. G., C. Luz, D. Mazzi, and T. C. M. Bakker. 2008. Inbreeding depression affects fertilization success and survival but not breeding coloration in threespine sticklebacks. Behaviour 145:425–441.

Gilk, S. E., I. A. Wang, C. L. Hoover, W. W. Smoker, and S. G. Taylor. 2004. Outbreeding depression in hybrids between spatially separated pink salmon, *Oncorhynchus gorbuscha*, populations: Marine survival, homing ability, and variability in family size. Environmental Biology of Fishes 69:287–297.

Goldberg, T. L., E. C. Grant, K. R. Inendino, T. W. Kassler, J. E. Claussen, and D. P. Philipp. 2005. Increased infectious disease susceptibility resulting from outbreeding depression. Conservation Biology 19:455–462.

Granath, K. L., W. W. Smoker, A. J. Gharrett, and J. J. Hard. 2004. Effects on embryo development time and survival of intercrossing three geographically separate populations of southeast Alaska coho salmon, *Oncorhynchus kisutch.* Environmental Biology of Fishes 69:299–306.

Guy, J. A., D. R. Jerry, and S. J. Rowland. 2009. Heterosis in fingerlings from a diallel cross between two wild strains of silver perch (*Bidyanus bidyanus*). Aquaculture Research 40:1291–1300.

Hughes, T. C., C. E. Lowie, and J. M. Haynes. 2005. Age, growth, relative abundance, and scuba capture of a new or recovering spawning population of lake sturgeon in the lower Niagara River, New York. North American Journal of Fisheries Management 25:1263–1272.

Kincaid, H. L. 1983. Inbreeding in fish populations used for aquaculture. Aquaculture 33:215–227.

Ludwig, A., N. M. Belfiore, C. Pitra, V. Svirsky, and I. Jenneckens. 2001. Genome duplication events and functional reduction of ploidy levels in sturgeon (Acipenser, Huso and Scaphirhynchus). Genetics 158: 1203–1215.

Maluwa, A. O., and B. Gjerde. 2006. Genetic evaluation of four strains of *Oreochromis shiranus* for harvest body weight in a diallel cross. Aquaculture 259:28–37.

McQuown, E., C. C. Krueger, H. L Kincaid, G. A. E. Gall, and B. May. 2003. Genetic comparison of lake sturgeon populations: Differentiation based on allelic frequencies at seven microsatellite loci. Journal of Great Lakes Research 29:3–13.

Miller, L. M., T. Close, and A. R. Kapuscinski. 2004. Lower fitness of hatchery and hybrid rainbow trout compared to naturalized populations in Lake Superior tributaries. Molecular Ecology 13:3379–3388.

Ryman, N. 1970. A genetic analysis of recapture frequencies of released young salmon (*Salmo salar* L.). Hereditas 5:159–160.

Schram, S. T., J. Lindgren, and L. M. Evrard. 1999. Reintroduction of lake sturgeon in the St. Louis River, western Lake Superior. North American Journal of Fisheries Management 19:815–823.

Welsh, A., M. Blumberg, and B. May. 2003. Identification of microsatellite loci in lake sturgeon, *Acipenser fulvescens,* and their variability in green sturgeon, *A. medirostris.* Molecular Ecology Notes 3:47–55.

Welsh, A., T. Hill, H. Quinlan, C. Robinson, and B. May. 2008. Genetic assessment of lake sturgeon population structure in the Laurentian Great Lakes. North American Journal of Fisheries Management 28:572–591.

Wright, S. 1978. Evolution and the genetics of populations: Variability within and among natural populations. University of Chicago Press.

LAURI KAY ELBING

Restoration and Renewal: A Sturgeon Tale

I CAN STILL REMEMBER THE DAY I FELL IN LOVE WITH THE LAKE STURGEON. I was passing through Tupelo, Mississippi, visiting my little sister in October 2001 on my way to the National Rails to Trails Conference in St. Louis, Missouri. She was very pregnant with my first niece, Sienna, who turned out to be unabashedly obsessed with fish, wildlife, nature, and science before she could even talk. Since it was my first time in Tupelo, I wanted to go exploring. We went on an adventure to see what we could see, and struck gold. Beyond being the birthplace of Elvis, Tupelo is home to the Private John Allen Fish Hatchery, the first National Fish Hatchery, established in 1901.

Among many past responsibilities, the Private John Allen Fish Hatchery has a restoration project to propagate three ancient species of fish that many biology geeks consider to be pretty cool: paddlefish, alligator gar, and lake sturgeon. My sister and I pulled up the driveway, passing an old antebellum home that once served as the hatchery manager's residence, now a gathering site for a local ladies social club. We went nosing around the ponds on site before seeing a couple men working in a giant outbuilding. It was filled with various sizes of what I can only think to call a holding tank, and looked like Sumo wrestler bathtubs. The gentlemen, dressed in unmistakable Fish and Wildlife green and brown, waved us in and began showing us and telling us about the fish they were working with as they moved some from a small tank to a larger one.

As the sturgeon outgrew their tanks they would become agitated, needing more space to swim. They would be moved up to sequentially larger environments, ultimately being introduced to the Tennessee River to live out their long lives.

One worker scooped up a tiny three-inch sturgeon and reached out his hand to give us a better look. That's when it happened. "Would you like to hold it?" he asked. "Oh my gosh, yes!" I replied immediately.

I was terrified; the little sturgeon was stunning to behold. The razor-sharp scutes were wrapped in soft shiny skin with striations in cool tones of rich soil. The fish wiggled and squirmed, then rolled over to reveal a bright white belly. Its tubular mouth protruded and sucked in the raw air, then contracted immediately into what appeared to be gray lips contorted into a silly little smile. I was hooked—so to speak—right then and there. What a fish.

I was familiar with the lake sturgeon, since I had heard about their plight in southeast Michigan during my tenure on the staff of Congressman John D. Dingell. We were in the process of establishing the first International Wildlife Refuge in North America, and sturgeon recovery was one of the justifications for the existence of the wildlife refuge.

Current sturgeon populations are about 1 percent of their former abundance, yet in recent years their story is also one of hope as they make their comeback in the Detroit River and Lake Erie region—with great thanks to Dr. Bruce Manny of the U.S. Geological Survey and Jim Boase of the U.S. Fish and Wildlife Service, among many others.

Almost everything I know about sturgeon is by way of these two amazing scientists. It is their collaborative spirit, political acumen, and dogged hard work, infused with an infectious enthusiasm, that have propelled the saga of the sturgeon into the consciousness of the communities and its leaders up and down the Huron-Erie corridor.

I find the sturgeon's ancient features elegant—even beautiful. From the youthful sharp scutes to the softened features of the aged, their silky skin reveals geometric structures under armor, adding texture and dimension to a delicate monochrome study in grayish sable. Leave it to a liberal arts sort like me to see beauty in the supposedly grotesque.

Lake sturgeon are an ancient species, genuine prehistoric relics and a keystone species that have thrived and survived unchanged for more than 100 million years, only becoming a threatened species in the Lakes Huron-Erie Corridor within the last 100 years or so. When the first Europeans made their way through Lake Erie and up through the strait now know as the Detroit River and finally into Lake Huron, the waters were teeming with sturgeon, especially during spawning season. Father Hennepin's account of spawning seasons brings to mind a sea boiling with giant fish, knocking and even capsizing canoes. He commented time and again in his journals about how plentiful and delicious the sturgeon were in Lake Erie and the Detroit River. This quote from Easter week in 1681 is as compelling as it is heartbreaking.

> After we had rowed a hundred Leagues all along the sides of the Lake Huron, we passed the Straights, which are thirty Leagues through, and the Lake of St. Claire, which is in the middle. Thence we arrived at the Lake Erie, or "of the Cat," where we stayed some time to kill Sturgeon, which come here in great numbers to cast their Spawn on the side of the Lake. We took nothing but the Belly of the fish, which is the most delicious part, and threw away the rest. (Hennepin 1903, 314)

In the early 1800s, lake sturgeon were still one of the most abundant fish species in the Great Lakes and the Huron-Erie Corridor served as the most important spawning and nursery habitat area in North America. Soon, sturgeon spawning habitats were being destroyed by the construction of dams and, adding to the pressure, the commercial value for smoked sturgeon and sturgeon caviar began to rise. With the growth of the nation came pollution from raw sewage and later from the many by-products of the industrial revolution and European agricultural practices of new inhabitants. The existence of sturgeon, like many other species, was threatened.

One major contributor to sturgeon decline was the cutting and dredging of channels as a means to support larger vessels and increased commerce; the most dramatic of these was the creation of the Livingston Channel.

In the 1910 *Economics of Lake Navigation*, the author argues the great value and savings Great Lakes navigation provides to the cost of basic needs and that from any investment the government makes in creating easier navigation, the entire nation will feel benefits in the price of bread. "The waterways are the lines of least resistance, and although Nature has not always provided deep channels through them, their economic possibilities are inestimable when all obstructions have been removed" (Anonymous 1910). What so many failed to recognize then, and often still do, is that we are a component of nature fully dependent on a complex system.

I have known since I was a little girl that we do not have a mutually dependent relationship with the natural world. It will go on without us, but without it, we will perish immediately. What we do ripples through this system in ways we don't realize or always recognize.

The lake sturgeon is considered an indicator species, according to an EPA initiative called the Detroit River–Western Lake Erie Basin Indicator Project (EPA 2010). An indicator is a measurable feature that provides useful information on ecosystem status, quality, or trends and the factors that affect them. The term "indicator" operationalizes a keystone species' health. The health of an indicator species lends insights into the health of its habitat. Lake sturgeon populations have never recovered from a century of overfishing, habitat loss, and pollution, but change is afoot, and hopeful signs are on the horizon for the sturgeon and for us.

For the first time in 30 years, lake sturgeon spawning was confirmed in the Detroit River in 2001. The site was located at Zug Island, where the fish spawned on coal cinder blocks, and while many thought that the spawning was likely unproductive, it was an encouraging event nonetheless. In response to that discovery, scientists conducted research to determine the extent of the sturgeon population in the Detroit River, including probable spawning locations, and conducted a few experiments. Beginning in 2004, reefs were constructed to increase available spawning habitat for lake sturgeon and other fish.

Michigan Sea Grant was the first to lead a consortium of federal, state, and private sector entities in the construction of three sturgeon spawning reefs in the waters off Belle Isle, the largest effort of its kind undertaken to date (Michigan Sea Grant 2005). The reefs were created out of limestone rock, cobblestone, and coal cinders, as the sturgeon had already shown a preference for these materials. Monitoring of these reefs continues, though they have not produced the coveted results. Disappointed but not discouraged, the project team built spawning reefs in other promising areas, including three Canadian sites: McKee Park in Windsor, Fighting Island in LaSalle, and Fort Malden in Amherstburg.

In October 2008 the United States and Canada partnered for the first time to construct a spawning reef near Fighting Island, owned by BASF Corporation. Within months, the reef was already being used by lake whitefish and other species, and by spring their number included lake sturgeon and the Northern madtom, an endangered species. Jim Boase noted the hopeful results: "Finding fertilized lake sturgeon eggs and collections of lake sturgeon larvae from the reef at Fighting Island this spring indicates that this restoration strategy is yielding very positive ecosystem results, and that this small, remnant population of native lake sturgeon may one day be restored to a higher level of abundance in the Detroit River" (Boase 2009, 7). One can easily see that the research findings point to former lake sturgeon spawning sites near Grassy Island, Sugar Island, and Sturgeon Bar as likely sites to be restored for spawning sturgeon (see the map at *http://www.epa.gov/ecopage/aquatic/lkstrugeon/fig1.pdf*).

I could be happy doing many different things professionally, but the most thrilling endeavors with which I have ever been involved centered on the Detroit River and

Lake Erie and required that I play a role in establishing the refuge, the conservation of Humbug Marsh, and the lake sturgeon recovery research and projects.

I met and talked with hundreds of people, with a steady 80 or so propelling restoration forward. I collaborated with many, fought with some, and outfoxed others under the steady tutelage of John Dingell, a monumental conservation hero. He always asked, "How's our river?" Like so many others, he was astonished and heartened to see the success of sturgeon habitat restoration. He said, "No one thought this degree of success was possible only 30 years ago." I've had so many conversations with people who grew up on the river and the lakes, watching the changes over time. Those who knew the river in the '40s, the '50s, the '60s, and even the '70s all say the same thing with goofy grins and twinkling eyes: "No one thought it was possible." Not in their lifetime—and while we aren't all the back yet, recovery is evidently under way.

In April 2006 I received an invitation to join Dr. Manny and Jim Boase as they collected data and samples from a site in Canadian waters just north of Fighting Island. The United States and Canada had just entered into a partnership to study recovery options and possible sites for another spawning reef, which could benefit a range of other species, including walleye, lake whitefish, and lake sturgeon. It was a great honor and thrill. I was going to hold a sturgeon again.

At the appointed time, on an ice-cold, gray, rainy-sleeting April morning, we met at the docks in Wyandotte, Michigan. It couldn't have been much above 32 degrees, but I never felt even a slight chill. Lindsey Haskins, a documentary film producer from California, was the other guest volunteer and took photos to document a day that I will treasure for the rest of my life. Haskins was developing a film on the lake sturgeon for the Public Broadcasting System, and we wore matching rain suits even though we did not call to coordinate our attire that morning. He was obviously a man of good taste and substance.

We loaded up the boat and headed out onto the Detroit River. As we passed Fighting Island, an eagle was being swarmed and attacked by herring gulls protecting their nesting grounds on the north side. You won't see that every day. We were off to a great start on our adventure to collect data, DNA, and memories.

Collecting biological information from each lake sturgeon when it is initially captured enables researchers to better understand the distribution and movement patterns, growth rates, and general health of a particular lake sturgeon when and if it is recaptured. More is better. The U.S. Fish and Wildlife Service began collaborating with several states and provinces by sharing information and data collected on sturgeon in 1994, an effort that now includes partnerships with 18 state, federal, and tribal agencies in the United States and Canada, eight universities, and over 25 nonprofit organizations, corporations, and commercial fishers (Boase 2009).

Finally, we pulled up to the first set of buoys, anchored and began by hauling in the gill nets to chart the other fish species that inhabit the waters near artificial reef sites. We counted the number of each species, then measured length and determined the gender and reproductive maturity of each of them. We brought in mostly walleye, as well as a few lake whitefish, bass, and even a little spotted gar.

Then the excitement began as we pulled in the setlines that were baited for lake sturgeon. We hauled several monsters into the boat, one at a time, and charted basic biological information from each fish, including length, girth, and weight. The next step, slicing off the leading edge of the pectoral fin ray, horrified me. While it is the equivalent of the Yakuza slicing off the pinky finger to signify gang membership, removing the fin ray is vital to the research and restoration efforts.

I was not prepared for what I saw, but recovered my composure and ignored my turning stomach as I learned the purpose. Age can be estimated by the fin ray, and tissue samples from the fin ray can be analyzed for DNA sequencing. We then tagged each of the creatures with two types of identification measures: a probe and a fin tag, with both the individual's own ID number and a data file. Set free once again in the river, they are gently lowered into the water and held there for just a minute or two until they regain orientation, then swim into the depths, and are gone.

The last sturgeon I brought in that day was the largest. Preparing to return to the docks, Jim Boase asked me to pull in the setline and remove the remaining bait. I did as I was instructed, until I felt that familiar tension and pulling on the line. I yelled back over my shoulder in excitement that we had another fish. Jim replied without looking up that this could not be the case.

He continued reviewing the data we had collected to that point. I raised my eyebrows and pulled on the line a bit more, feeling the fish swimming back and forth slowly, then under the boat. I yelled back at Jim again that I had another fish, to which he replied, "No, you don't. That's just the current." I then yelled back over my shoulder a third time, "I have a fish! I know the difference between a fish and the current. I've been fishing since before I could walk, so get the net!"

That got his attention.

He peered over the edge of the boat, still not completely believing me, but to his great surprise, I had another fish. He rushed over, grabbed the net and we brought her in. What a gorgeous creature.

When our work was done we motored back to the docks in Wyandotte. Ice formed and collected on the windshield as we pressed on, but I felt only warmth and exhilaration. It would last for weeks. I thought about the great spans of time when these giant creatures thrived and what it must have been like for the people that witnessed the spectacle of spawning season in the straits before the sturgeon were wiped out. I thought about the passion, intelligence, creativity, and savvy of scientists like Manny and Boase and was grateful to them for so many things: for the work they do so well, for their friendship, and for giving me one of the most exquisite gifts of my life in that day out on the river.

The first time I held that tiny little sturgeon in Tupelo was a month after 9/11. The Rails to Trails conference in St. Louis was almost canceled, but we all decided to persevere—be it in defiance or the need to stay busy and stay on task. It was a time when many Americans were looking for reassurance and peace of mind. The director of the National Park Service was the keynote speaker, and she talked of a surge in visitors to our parks. We were drawn to parks and nature areas in droves in the days following 9/11. Green space across the nation was crowded with people seeking solace and comfort, escape from the incessant news and pictures, with some unsure what drew them to those places. There are ample studies that show the psychological and physical benefits we gain through exposure to nature. That tiny little sturgeon with the ancient history, one that predates even the dinosaurs, and a future of restoration and renewal offered me consolation and reassurance that even in the face of devastation there is hope for them, and there is hope for us. Life goes on.

REFERENCES

Anonymous. 1910. Economics of lake navigation. Old and Sold Antiques Digest Http://www. oldandsold.com/articles36/inland-seas-23.shtml.

Boase, J. 2009. Lake sturgeon have arrived. Fisheries and Aquatic Resources Program, Region 3, U.S. Fish and Wildlife Service, Ft. Snelling, Minnesota. Fish Lines, June, 6–7.

EPA (U.S. Environmental Protection Agency, Large Lakes and Rivers Forecasting Research Branch). 2010. Detroit River–Western Lake Erie Basin Indicator Project. Http://www.epa. gov/med/grosseile_site/indicators/index.html.

Hennepin, L. 1903. A new discovery of a vast country in America. Reuben Gold Thwaites, ed. Vol. 1. McClurg. Http://www.archive.org/details/newdiscoveryofva02henn.

Michigan Sea Grant. 2005. Lake sturgeon habitat protection background. Http://www. miseagrant.umich.edu/sturgeon/background.html.

PIERRE DUMONT AND YVES MAILHOT

The St. Lawrence River Lake Sturgeon: Management in Quebec, 1940s–2000s

THE LAKE STURGEON (*ACIPENSER FULVESCENS*) EASILY DISTINGUISHES ITSELF from other fish in many aspects: its particular physical appearance, the time frame of its life cycle, and the historic importance, veneration, and respect given it by the First Nations. Another very distinctive characteristic is the similarity between the lake sturgeon's and the human life cycle: contrary to the other fish species, whose life cycle is short or at least much shorter, the number of years needed to produce a generation is about the same for both, and 100-year-old specimens could still be fished or sampled some decades ago in the St. Lawrence River. Accordingly, acquiring the useful scientific knowledge and experience to define and support management is also a matter of generations, but in this case, of scientists.

The first data-collecting period in the St. Lawrence River covered the 1940s to 1970s. In these years, studies were mainly descriptive of the biological characteristics and the movements of lake sturgeon. Commercial fishermen were still catching very large specimens, and few fishery management restrictions were implemented. Concerns about the sustainability of the fishery and habitat protection were expressed in the 1960s and 1970s. A second data-collecting period (1980s–2000s) was realized by a third generation of scientists and involved more precise and detailed population dynamics studies, since an overexploitation diagnosis was confirmed in 1987, after an increase of fishing pressure (Dumont, Axelsen, et al. 1987). Restrictions were then applied to the sport and commercial fisheries but were not found sufficient to reverse the decline.

In 2000, a stringent management plan designed to adapt the fishing catch to the natural production capacity of the downstream lake sturgeon population rapidly produced positive reactions in this sturgeon population, the only one in the lower part of the St. Lawrence system for which linear fluvial habitat integrity has nearly been maintained to the present. In Quebec, in the three other St. Lawrence system populations, Ottawa River, Lac Saint-François, and Lac des Deux Montagnes (plate 10), major constraints occurred (mostly habitat fragmentation and severe pollution episodes) that reduced very significantly the sturgeon production, as will be detailed further in this chapter.

We will resume successively the most important information relative to the perception and use of lake sturgeon by First Nations and first European settlers, the two scientific data-collecting periods between the 1940s and the present and their respective management actions, and finally, conclude on the challenge created by the need to continue to collect scientific information in order to permit sound management decisions to maintain the production capacity for the future generations of sturgeon, man, fishers, and . . . scientists.

Food and Spirituality for the First Nations

It is not possible to find the origin of the interest of the different First Nations' tribes in lake sturgeon, since no written documents exist from so long ago, but the first French explorers testified that they had a great interest in the fish. Historical use of lake sturgeon by First Nations has also been well documented by archeological observations in the Quebec distribution area. For example, according to Clermont, Chapdelaine, and Cinq-Mars 2003, the remains of 70 lake sturgeon found in Morrison Island and 105 specimens in Allumette Island, on the Ottawa River, strongly suggest that they were also important food items for Native people of the Laurentian Archaic period (6,100–5,500 BP). In the site of Pointe-du-Buisson, at the confluence of the St. Lawrence River in Lac Saint-Louis (plate 10), bone fragments of lake sturgeon, channel catfish, and catostomids, among the most fatty species of the St. Lawrence River fish community, consistently dominate in the remains of meal identified, associated with the Middle Late Woodland period (1,450–1,000 BP) (Courtemanche 2003). The Mohawk people of Kahnawake are still practicing subsistence fishery in Lachine rapids, near Montreal, at the outlet of Lac Saint-Louis.

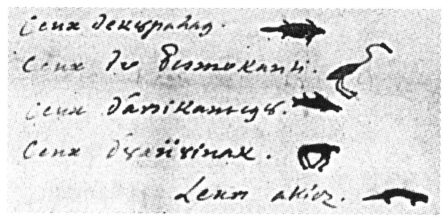

Figure 1. Animal totems used in the 1700s as signatures by several Abenakis and other tribes. Each family band had a sacred totem that symbolized the band's spiritual closeness with a specific creature [Callaway 1989]. Odanak Abenakis signature of sturgeon is central in photo.

Mélançon (1936) reports that sturgeon fishing is very ancient in Canada and that the First Nations fished with coarse nets during the wintertime and with harpoons in summer, and that they were catching large quantities in Trois-Rivières. He also noted that sturgeon was the ritual food for the Algonquin's wedding meals and that the different Algonquin tribes such as the Abenakis, the Mahicans, and the Penobscots were nicknamed "The Sturgeons" by the Iroquois nation.

The Odanak Abenakis were also known to use a drawing of the shape of a sturgeon as a seal to sign treaties with the Europeans (figure 1), and today, the sturgeon appears on the armorial bearings of their Council (plate 11a), which was also proudly represented as a beautiful vegetal sculpture in 2000 and 2001 within the Mosaicultures in Montreal, a public large-scale special exposition in which the First Nations participated (plate 11b). Furthermore, when this nation, settled along the Saint-François and the Bécancour rivers (plate 10), concluded an official agreement with the government of Quebec in 2000 relative to the maintenance of particular privileges, the sturgeon was clearly attributed a sacred special status.

The Abenakis are not the only Quebec First Nation to bear a particular reverence to the lake sturgeon. Sylvie Beaudet, a biologist colleague involved with the Cree nation in Waswanipi (in the James Bay watershed), observed a special attitude in people working in a small fish-processing plant. According to a local belief, only the best fishermen may reincarnate as a sturgeon and, consequently, the processing of sturgeon carcasses is done at a very slow pace and with respect and meditation.

Another colleague biologist representative of the Cree Council, René Dion, noted that fishermen will attribute a special personality to this fish and mentioned that they catch other species of fish, but that they kill a sturgeon.

Finally, another point of importance worth mentioning lies in the fact that when the James Bay Territory Convention was signed in 1976, this official agreement between the Cree and Inuit nations and the governments of Quebec and Canada relative to the hydroelectric development of northern Quebec, the lake sturgeon exploitation was totally reserved to the exclusive use of people from the Cree and Inuit nations, considering the great symbolic and dietary importance of this fish to Native people.

Food and Privilege for the European Settlers

The first European settlers were readily interested in lake sturgeon as a food source. Pierre Boucher, French governor of Trois-Rivières, reported in 1664 that lake sturgeon was abundant upstream of Quebec city, especially in all the St. Lawrence fluvial lakes and at the actual Montreal city site, and that these fish were very large, mostly four, six, and eight feet long (Boucher 1664). He also noted that salted sturgeon meat was as good after two years of conservation as four days after the catch.

Samuel de Champlain (cited by Giguère 1973), known as the founder of Quebec city, also mentioned in 1613 the (Lachine) rapids near Montreal as a place where sturgeon were caught. André-Napoléon Montpetit, in his 1897 book in French on the freshwater fishes of Canada, described sturgeon fishing in detail as an activity mostly occurring at night, in which a long wood pole equipped with a large iron hook was used. He also stated that according to ancient French and English laws, the local Canadian governors had the privilege of claiming a special annual reserve of sturgeon meat.

Three Generations of Scientists for Three Generations of Lake Sturgeon

Most lake sturgeon populations in the St. Lawrence River–Great Lakes system declined rapidly at the turn of the twentieth century, and it is recognized that overfishing and habitat degradation have both contributed to this decline. The result has most often been the same, a relatively high initial yield, followed by a sudden and permanent decline to a very low level (Harkness and Dymond 1961; Scott and Crossman 1973). From 1920 to 1980, with relatively high and stable annual catches

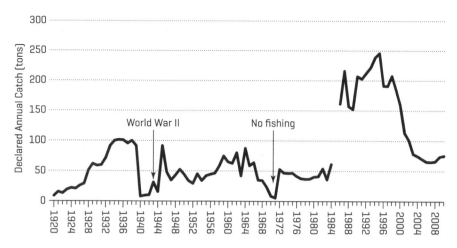

Figure 2. Evolution of declared lake sturgeon commercial landings in the Quebec part of the St. Lawrence River, 1920–2007 (Robitaille et al. 1988).

averaging 50 metric tons and corresponding approximately to a 0.5 kg/ha yield, the St. Lawrence River lake sturgeon fishery did not follow the general and distressing pattern observed upstream (figure 2).

Historical factors explaining this relative stability are not well understood, but are probably related to the large size of the river and the productivity of its diverse nonfragmented habitats along a 350 km stretch. Fishing pressure and poaching were likely high. In the 1920s more than 2,000 commercial fishing licenses were issued in the Quebec part of the St. Lawrence River; this number gradually decreased to around 1,000 in the 1950s (Robitaille et al. 1988), while it is now 134, of which only 52 are authorized for lake sturgeon fishing. Available information also suggests that the price offered to fishermen was high, and sometimes higher than current value.

For example, in the 1940s, according to Vladykov and Beaulieu (1946), commercial fishermen were receiving between 60 cents and one dollar per pound of lake sturgeon flesh, a price equivalent to $17–$28 per kilogram today. However, the lower St. Lawrence sturgeon fishery was far from the Great Lakes freshwater market, and it does not seem that the caviar industry experienced the dramatic development observed upstream. Concerns about the necessity of conservative measures appeared early in the available historical sources. For example, Montpetit (1897) proposed the application of a 3.5 feet (107 cm) minimum size limit to the fishery. We do not know if this specific size limit was applied at the beginning of the twentieth century, but a minimal size of 45 cm of dressed length (approximately 80 cm total length) was enforced in the 1940s (Cuerrier and Roussow 1951) and was likely implemented earlier, in 1927 (Harkness and Dymond 1961).

This minimum length, still currently applied, is clearly not long enough to protect the spawning stock, but it is the base of the historical widespread use of the 19 to 20 cm stretched mesh gill net, a gear that mainly selects lake sturgeon from 80 to 120 cm (Dumont, Fortin, et al. 1987; Fortin, Guénette, and Dumont 1992). This gill net being the major gear used, the fishery likely evolved over a long period under a combined minimum/maximum size limit. When the exploitation rate was not too high, this allowed at least partial protection to the spawning stock, composed of long-lived and highly fecund multiple spawners. While very large specimens (> 100 kg) like those reported by Vladykov (1955) or Harkness and Dymond (1961) are now very rare or completely absent in the lower St. Lawrence River, fish exceeding 40 kg and 1.8 m are still observed in the spawning grounds (Fortin, D'Amours, and Thibodeau 2002; Dumont et al. 2011) and in the commercial catch. In the past 25 years, maximum age and weight reported in the commercial harvest were 96 years and 90 kg (Dumont, Fortin, et al. 1987).

In the 1910s and 1920s, Quebec sport fishing rules imposed a closure period in June, and it seems that this measure was also applied to commercial fishing. In 1950, this protection moved to the month of May, in 1956, to a mid-May to mid-June period, and in 1967, to the beginning of the ice-over period (generally in December) to the end of April. A bag limit of two sturgeons appeared for the first time in 1971, simultaneously with a return to the mid-May to mid-June closure period. This period was extended in 1984 (mid-April to mid-June) and the bag limit reduced to one fish in 1988. The current period extends from mid-June to the end of October, with a closure period between August 1 and September 14 for the commercial catch in order to reduce incidental mortality during the warmer season.

Another factor contributing to the relative stability of the lake sturgeon fishery, at least for the second half of the twentieth century, is the early development, from 1941, and use for management purposes, of scientific knowledge of the species' biology in the Quebec part of the St. Lawrence and the Ottawa rivers watersheds under the patronage of the Station biologique de Montréal. The first research works were realized in cooperation with commercial fishermen, some of them being qualified as excellent naturalists (figure 3). Within 10 to 20 years, different aspects have been investigated : morphology and systematics (Vladykov and Beaulieu 1946; 1951; Roussow 1955a; Vladykov and Greeley 1963; Magnin 1962), feeding (Fry et al. 1941; Vladykov and Gauthier 1941; Vladykov 1948), movements of juvenile fish (Roussow 1955b; Magnin and Beaulieu 1960), growth (Cuerrier and Roussow 1951; Roussow 1955b; Magnin 1962), sexual maturation, spawning periodicity, and gonad development (Cuerrier 1945; Dubreuil and Cuerrier 1950; Roussow 1957, Cuerrier 1966), and commercial fishing (Cuerrier, Fry, and Préfontaine 1946; Roussow 1955b; Cuerrier 1962). With the exception of some concerns about dams' impacts on sturgeon movements between Lac Saint-François and Lac Saint-Louis (Roussow 1955b), and some details about spawning grounds location in the Saint-François (Cuerrier 1962), Batiscan and Chaudière rivers (Vladykov 1955), little attention was paid to habitat before the work of Mongeau, Leclerc, and Brisebois (1982) in the 1960s and 1970s on the natural restoration of the sturgeon population in Lac des Deux Montagnes following an anoxic period in the early 1950s.

A second phase of development of scientific knowledge on lake sturgeon biology and exploitation was initiated in 1981, when an increase in demand for commercial fishing licenses led us to study various aspects of the biology, habitat, and dynamics of this population, to revise the fishery management plan, and to implement various measures of habitat protection and improvement. This work has been done in close collaboration with the Department of Biological Sciences of the Université du Québec à Montréal.

Lake Sturgeon Biology in the Lower St. Lawrence River System

DISTRIBUTION AND STOCK COMPOSITION

Lake sturgeon occur all along the Quebec part of the St. Lawrence River, where the species now forms two populations. The first one, located in Lac Saint-François, has been gradually separated from the downstream and upstream groups by the construction of the Beauharnois–Les Cèdres (1912–1961) and the Moses-Saunders (1958) hydropower complexes (plate 10). Tagging studies in the 1940s (Roussow 1955b) indicated that lake sturgeon were then able to migrate along the St. Lawrence River, from the limits of the brackish waters up to at least Ash Island, near the outlet of Lake Ontario (figure 4).

Downstream Lac Saint-François, over a 350 km stretch from Beauharnois Dam, at the head of Lac Saint-Louis, to the brackish waters downstream of Quebec city (plate 10), lake sturgeon likely form a homogeneous phenotypic and genotypic stock (Guénette, Rassart, and Fortin 1992; Guénette, Fortin, and Rassart 1993). In the freshwaters of the upper estuary, downstream of Lac Saint-Pierre, this population co-occurs with the Atlantic sturgeon (*A. oxyrynchus*). This lake sturgeon population was not found to be significantly different from the population of Lac des Deux Montagnes, an enlargement of the lower Ottawa River downstream from Carillon Dam. However, the presence of a particular genotype in Lac des Deux Montagnes, and its absence in the rest of the St. Lawrence, support the hypothesis that sturgeon

Figure 3. Mr. Napoléon Lalumière, a well-known and curious Lac Saint-Louis lake sturgeon commercial fisherman, with a girlfriend circa 1940. Mr. Lalumière also contributed to the identification of a new species of catostomid, the copper redhorse (*Moxostoma hubbsi*), an endangered species exclusive to southwestern Quebec.

movements between the two water bodies are limited, as shown by the results of tagging studies (Fortin et al. 1993) and morphological comparisons (Guénette, Rassart, and Fortin 1992).

Mitochondrial DNA variation also suggests that genetic heterogeneity seemed lower in the St. Lawrence River population than in the James Bay drainage population, likely because this southern population has been more significantly influenced by overfishing and man-made habitat changes over a long period of time (Guénette, Fortin, and Rassart 1993). Small highly fragmented lake sturgeon groups also occur upstream of Lac des Deux Montagnes along the Ottawa River, where, in the last 580 km, nine reaches were gradually separated by dams during the nineteenth and twentieth centuries (Haxton and Findlay 2008).

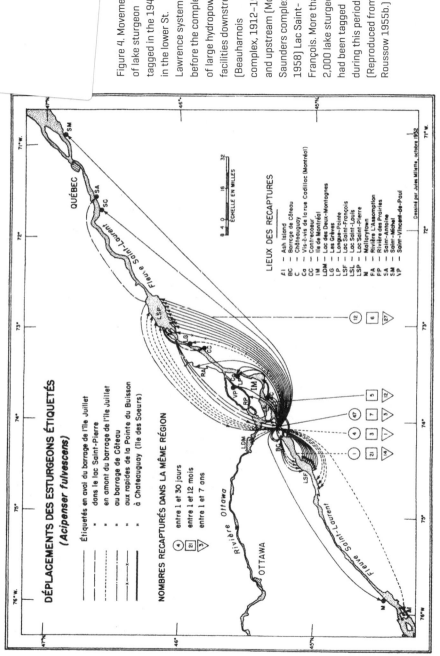

Figure 4. Movements of lake sturgeon tagged in the 1940s in the lower St. Lawrence system before the completion of large hydropower facilities downstream [Beauharnois complex, 1912–1961] and upstream [Moses Saunders complex, 1958] Lac Saint-François. More than 2,000 lake sturgeon had been tagged during this period. [Reproduced from Roussow 1955b.]

DÉPLACEMENTS DES ESTURGEONS ÉTIQUETÉS
(Acipenser fulvescens)

Étiquetés en aval du barrage de l'Île Juillet
dans le lac Saint-Pierre
" " en amont du barrage de l'Île Juillet
au barrage de Côteau
" " aux rapides de la Pointe du Buisson
à Chateauguay (Île des Soeurs)

NOMBRES RECAPTURÉS DANS LA MÊME RÉGION

⊙ entre 1 et 30 jours
☐ entre 1 et 12 mois
▷ entre 1 et 7 ans

LIEUX DES RECAPTURES

AI — Ash Island
BC — Barrage de Côteau
C — Chateauguay
Ca — Vis-à-vis de la rue Cadillac (Montréal)
CC — Contrecoeur
IM — Île de Montréal
LDM — Lac des Deux-Montagnes
LG — Les Grèves
LP — Longue-Pointe
LSF — Lac Saint-François
LSL — Lac Saint-Louis
LSP — Lac Saint-Pierre
M — Mallorytown
RA — Rivière L'Assomption
RP — Rivière des Prairies
SA — Saint-Antoine
SM — Saint-Michel
VP — Saint-Vincent-de-Paul

ÉCHELLE EN MILLES
8 4 0   8   16   32

Dessiné par Jules Millette, octobre 1952

QUÉBEC

Fleuve Saint-Laurent

Rivière Ottawa

OTTAWA

Fleuve Saint-Laurent

Recent assessment of lake sturgeon population genetic structure in the Great Lakes basin and Hudson Bay drainage (Welsh et al. 2008), using 27 spawning locations, indicated that, even if the majority of the 25 spawning populations within the system were genetically distinct from each other, the Lac Saint-François population (represented by the Grasse River sample) and the Lower St. Lawrence River population (downstream from Beauharnois Dam, represented by the des Prairies River sample) were grouped with neighboring populations in the upper St. Lawrence River and in Lake Champlain. Before the construction of two dams on the Richelieu River in the 1840s, at Chambly and St-Ours, there was no obstacle to fish passage between Lake Champlain and the St. Lawrence River (Dumont et al. 1997).

MOVEMENTS

In the Quebec part of the St. Lawrence River, mark-recapture experiments on more than 12,000 lake sturgeon indicate that, with the exception of spawning migrations, which are extensive, movements are restricted (Roussow 1955b; Magnin and Beaulieu 1960; Dumont, Fortin, et al. 1987; Fortin et al. 1993). Most multiple recaptures (up to four) were done near the tagging sites. Globally, juvenile sturgeon were found sedentary, but some movements were observed on a small proportion of them (<5 percent). Larger sturgeon (>85 cm), tagged on the des Prairies and l'Assomption rivers spawning grounds, and on pre- and postspawning concentration sites, were recaptured throughout the St. Lawrence River, from Beauharnois to the upper estuary (Dumont, Fortin, et al. 1987; Fortin et al. 1993).

In this system, lake sturgeon occur in large numbers in small localized sites, increasing their vulnerability to fishing gear and to any intervention on these local habitats (filling, dredging, toxic outflows, etc.). Some sturgeon seem to form very stable groups; for example, at least on three occasions, pairs of fish tagged simultaneously were recaptured together (Dumont, Fortin, et al. 1987).

Larvae drift downstream from the main spawning ground (des Prairies River), and size and age distribution of juveniles in the experimental samples (mostly age two to eight), and of subadults in the commercial harvest samples, suggest that sturgeon are mainly produced in the major spawning grounds, in the upper part of the system. Their offspring would then quickly drift downstream to the upper estuary and later, slowly and gradually colonize the river along a downstream-upstream gradient. Most juvenile sturgeon concentrations are found in the lower part, between the Lac Saint-Pierre archipelago freshwaters and the estuarine brackish waters, near Orléans Island (figure 5). In the commercial catch of the upper estuary (Gentilly), males and females are smaller, lighter, and younger, and only a small proportion of these fishes

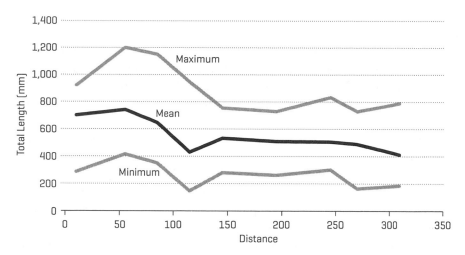

Figure 5. Average size of juvenile lake sturgeon in the 1992–1999 experimental multimesh gillnet catch from Lac Saint-Louis to the upper estuary. [Adapted from Dumont et al. 2000.]

are maturing (table 1), while upstream in Lac Saint-Louis, they are longer, heavier, and older, and almost half the females are maturing. Intermediate values are generally observed in Lac Saint-Pierre and its archipelago.

GROWTH

Lake sturgeon average back-calculated total lengths at age 5, 10, 15, 20, 25, 30, 35, and 40 years in the lower St. Lawrence River are respectively 530; 770; 977; 1,165; 1,251; 1,338; 1,391; and 1,496 mm (Fortin, Guénette, and Dumont 1992; see also Fortin et al. 1993). Average corresponding weights are 0.8, 2.8, 6.0, 10.7, 13.6, 16.9, 19.2, and 24.4 kg. Condition factor (a function of length and weight that ranges from 0 to 1) increases with size (from .51 to .83 for 500 to 1,500 mm fish) and, at equal size, is higher in the St. Lawrence River than in the Lac des Deux Montagnes (Fortin, Guénette, and Dumont 1992; Guénette, Rassart, and Fortin 1992).

In the river, as observed in the whole distribution area, growth rate decreases along a latitudinal gradient (Fortin, Dumont, and Guénette 1996). Compared to other eastern populations (≤ 80°W), lake sturgeon grow relatively rapidly in the St. Lawrence River; however, growth is faster in Wisconsin inland waters and Lake of the Woods, in the western part of the distribution area (Fortin, Dumont, and Guénette 1996).

FOOD AND FEEDING

Juvenile lake sturgeon are generalists and opportunistic benthic feeders. In the lower St. Lawrence River, their diet has been found to be highly diversified and composed of at least 75 taxa, of which more than 50 occurred in more than 5 percent of the samples. This probably reflects the high diversity of the benthic fauna of this system, which is also much more productive (~2400 invertebrates/m² compared for example to <100 invertebrates/m² in northern Ontario watersheds (Nilo et al. 2007; Beamish, Noakes and Rossiter 1998). Food preferences observed vary according to site, period, and fish length. Their prey are small, and the most exploited groups are amphipods, aquatic insect larvae, mollusks, and oligochaetes. Fish and microcrustaceans are also eaten, but in much smaller proportions (Nilo et al. 2007; Mongeau, Leclerc, and Brisebois 1982). In the Nilo et al. (2007) study, diet composition was only partly determined by benthos availability, and there was a positive selection for drifting prey. In the St. Lawrence River, lake sturgeon aggregate locally, but the presence of juveniles in these specific areas cannot be fully explained by their food habits.

In the upper estuary, in the co-occurrence zone, Atlantic and lake sturgeon eat the same major prey in different proportions. Guilbard et al. (2007) observed that age-0+ fish of both species fed mainly on the crustacean in the Amphipod family Gammaridae. Juvenile and subadults of both species fed mainly on oligochaetes and gammarids, but in opposite proportion: gammarids were the dominant prey for lake sturgeon and oligochaetes for Atlantic sturgeon. In addition to gammarids, the lake sturgeon diet included insect larvae and mollusks, whose proportions increased with fish size. The fish contribution to the summer diet was higher in the estuary than in the riverine section of the river. A comparative morphometric study of the digestive tract suggests that the thicker gizzard wall of the lake sturgeon facilitates the crushing of hard prey, while, for Atlantic sturgeon, a longer intestine and a higher development of the spiral valve favor chemical digestion (Guilbard et al. 2007).

Fresh and decomposing vegetation was only occasionally observed in the stomach contents of juvenile lake sturgeon in the estuary (Guilbard et al. 2007), but was observed much more frequently upstream, in the riverine section (Nilo et al. 2007). Dreissenid mussels (*Dreissena polymorpha* and *D. rostriformis*) are now found in the digestive tract of lake sturgeon but do not seem to be selected (Nilo et al. 2007; Guilbard et al. 2007; Pierre Dumont, unpublished data).

REPRODUCTION AND RECRUITMENT

In lake sturgeon sexual maturity is delayed, the female median age at first maturity being 26 years (Guénette, Rassart, and Fortin 1992). Spawning periodicity is estimated to be one to three years for males and is likely higher than four years for females (Fortin, D'Amours, and Thibodeau 2002). Spawning grounds are generally located in the St. Lawrence tributaries, in the lower reaches of des Prairies, des Mille Iles, l'Assomption, Ouareau, Saint-François, Saint-Maurice, Batiscan, and Chaudière rivers, downstream from natural falls or dams (plate 10). However, in 2002, a spawning ground of 35,000 m² , the largest known in the river, was discovered in the Lachine rapids (La Haye et al. 2003; 2004) at the outlet of Lac Saint-Louis, one of the only two large rapid sectors that has not been impounded along the St. Lawrence River. Lake sturgeon also spawn in the residual bed of the St. Lawrence River, at its inlet into Lac Saint-Louis, in a sector of rapids highly utilized for fish reproduction (Montpetit 1897) before the progressive derivation of more than 85 percent of the river discharge in the Beauharnois Canal between 1929 and 1961 (plate 10).

Reproduction occurs during the second to the fourth week of May in the tributaries (12–17°C) and later in the St. Lawrence River, between the fourth week of May and the third week of June (11–14°C), because of the slower warming of the highly transparent water flowing from the Great Lakes. In des Prairies River, two peaks of spawning activity (as measured by sturgeon catch per unit of effort and egg deposition) were observed each year. This bimodality was not related to water temperature or to river discharge, and recapture data suggest some fidelity of lake sturgeon to the first or second spawning period. Sturgeon caught during the first period were generally larger than those of the second peak (Fortin, D'Amours, and Thibodeau 2002; Dumont et al. 2011).

Spawning beds are covered with a mix ranging in size from fine to medium-sized gravel to boulders, under 0.1 to more than 6 m of water and exposed to current velocities between 0.1 and 1.9 m/s. In des Prairies River, larval drift generally occurs at night and lasts 14 to 20 days, between the third week of May and the third week of June (La Haye et al. 1992; 2003; 2004; Fortin, D'Amours, and Thibodeau 2002; D'Amours, Thibodeau, and Fortin 2001; Garceau and Bilodeau 2004). On the des Prairies River spawning grounds, egg predation by other species (240 fishes belonging to nine species) or by lake sturgeon (31 fishes tested by gastric lavage) has been found to be of very low (<1 percent) occurrence (Fortin, D'Amours, and Thibodeau 2002). However, these observations do not apply to small-size fishes, including a new highly invasive species now well established in the lower St. Lawrence system, the round goby (*Neogobius melanostomus*).

Fecundity has been measured on 16 females caught in Quebec from the 1940s to the 1970s (figure 6). In this data set, the number of eggs per female varies between 48,420 and 670,450 for females from 857 to 1,854 mm and from 5.1 to 52.7 kg, corresponding to a relative fecundity of 13,200 eggs/kg.

Year-class strength appears to be determined in the first few months of life.

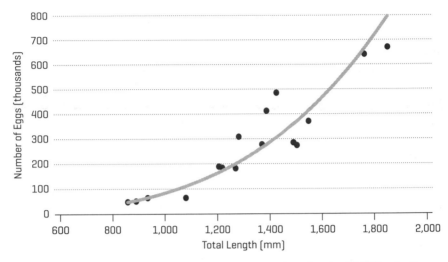

Figure 6. Lake sturgeon fecundity in Quebec waters. Data are from Cuerrier [1966] for the St. Lawrence River [*n* = 11], Dubreuil and Cuerrier [1950] for the lower Ottawa River [*n* = 4], and from Magnin [1977] for the La Grande River in the James Bay watershed [*n* = 1].

Climatic (daily rate of increase of water temperature in May and June) and hydrological conditions in June (water discharge of the des Prairies River), during which larvae drift from spawning grounds and exogenous feeding begins, were identified as critical determinants of year-class strength in the St. Lawrence River (Nilo, Dumont, and Fortin 1997, Dumont et al. 2011).

Anthropogenic Limiting Factors

HABITAT FRAGMENTATION

Recent experimental fishing by the Ministère des Ressources naturelles et de la Faune du Québec (MRNF) and by the Ontario Ministry of Natural Resources (OMNR) confirm that the Lac Saint-François sturgeon population, upstream from Lac Saint-Louis, considered as depleted in the 1940s (Cuerrier and Roussow 1951), the 1960s (Joliff and Eckert 1971), and the 1980s (Dumont, Fortin, et al. 1987), is still of very low abundance. In the 1940s, lake sturgeon were still able to move along the St. Lawrence River, from the limits of the brackish waters up to at least the outlet of Lake Ontario (Roussow 1955b). Depletion of the Lac Saint-François lake sturgeon population can likely be related to the combined effects of the progressive construction of dams at both extremities of the lake between 1912 and 1961 (Morin and Leclerc 1998) and of the overfishing of the residual stock. Sturgeon fishing has been prohibited since 1987.

Crib dams were constructed in the 1870s along the Ottawa River to facilitate log drives, forming the first true barriers to upstream movement of lake sturgeon. Small hydroelectric dams were first constructed in the 1880s, and large-scale hydroelectric construction began in 1948. According to Haxton (2008), these dams had the greatest impact on sturgeon through fragmentation of the habitat and flooding of natural rapids and shore lands. Since 1964, sturgeon movement between the St. Lawrence River and the lower reach of the Ottawa River has been almost completely blocked by the Carillon hydroelectric dam at the head of Lac des Deux Montagnes. Haxton and Findlay (2008) observed significant variation in relative sturgeon abundance among nine Ottawa River reaches with the greatest abundance in unimpounded reaches and concluded that water power management is the primary factor affecting lake sturgeon in this river.

WATER POLLUTION

The Lac des Deux Montagnes sturgeon population, which has not been exploited commercially since 1950, has been decimated following a prolonged winter period of anoxia caused by the discharge of untreated domestic and pulp and paper effluents in the Ottawa River watershed. This drastic reduction acted as a population bottleneck. Surveys in 1964, 1965, 1966, 1968, 1969, 1978, and 1979 showed that the restoration of this population was very slow and, after almost 30 years, was still quite fragmentary, as only seven year classes were identified in the experimental catch. Two (or three) year classes were dominant (1955 and 1960–1961) and no individual born before 1951 was observed (Mongeau, Leclerc, and Brisebois 1982). Only one additional year class was detected in a 1985 survey (MRNF, unpublished data).

In the L'Assomption River, a 10-fold improvement in larval production was observed following water treatment installation, while massive egg and larvae mortality occurred before sewage treatment (Dumas, Trépanier, and Simoneau 2003). In an attempt to evaluate the effects of contaminants on lake sturgeon in the 1990s, bioindicators were measured in a sample of lake sturgeon from two sites: des Prairies River and a reference site in the upper reach of the Ottawa River in the La Vérendrye Wildlife Reserve. Negative effects of organic contaminants, including high levels of PCBs, were suspected, fish taken from des Prairies River having moderate to severe hepatic (liver) pathology (Rousseaux, Branchaud, and Spear 1995).

According to Doyon, Fortin, and Spear (1999), the prevalence of fin deformities among larvae raised in artificial streams was significantly greater in the progeny of sturgeon sampled in des Prairies River (6.3 percent) compared with the progeny of sturgeon from the reference site (1.7 percent). Juvenile sturgeon from the St. Lawrence River and tributaries were estimated to have a 2.9 percent prevalence of fin and craniofacial malformations. Concentrations of liver and intestine reserves of an active form of vitamin A were also found to be significantly lower (as much as 40 times lower) in the des Prairies River sample (Doyon et al. 1998; Doyon, Fortin, and Spear 1999; Ndayibagira et al. 1995). Excessive and insufficient concentrations of this retinoid elicit development anomalies. Veillette (2007) also reported a higher prevalence of histological anomalies in the liver of sturgeon sampled in Lac Saint-Louis than in a pristine area (Lac Berthelot, James Bay drainage); these anomalies were generally related to high values of the hepatosomatic index (>3.15 percent).

DREDGING

Sediment disposal operations can affect sturgeon by causing direct mortality of different life stages by burial and gill clogging, or indirectly by habitat degradation. Since 1844, dredging and dumping operations conducted for the progressive construction of the navigation channel (between Quebec City and Montreal: current minimum width 244 m, and minimum depth 11.3 m) and of the St. Lawrence Seaway (between Montreal and Lake Ontario: current minimum width 24 m, and minimum depth 8.2 m) caused major man-made habitat changes. Recent trawl surveys in Lac Saint-Pierre and the upper estuary (MRNF, unpublished data) indicate that the navigation channel is a poor-quality habitat for fish, likely because of high current velocity and frequent ship passage, but that some deepwater habitats alongside the channel are very important for lake sturgeon, acting all year long as refuges and feeding grounds. More than 175 million m³ of sediments were dredged and dumped in the river during the twentieth century, and channel and harbor maintenance is of annual occurrence (Dumont, Fortin, et al. 1987; Robitaille et al. 1988). In order to protect sturgeon habitat in the river, there is a need to acquire a deeper knowledge about the relationship between habitat characteristics, feeding, and distribution of lake sturgeon at all stages of the life cycle (Nilo et al. 2007). In a large-scale study on sediment disposal in the St. Lawrence upper estuary, Hatin, Lachance, and Fournier (2007) observed site avoidance and negative impact of sediments disposal operations for Atlantic sturgeon but not for lake sturgeon, likely because the lake sturgeon diet is more diversified.

FISHERY

In southern Quebec, lake sturgeon is mainly a commercial species. In 1987, the St. Lawrence stock was considered overexploited because of high annual natural and fishing mortality rates of the exploited segment (ages 15 to 30), unbalanced age structure, deficit of reproductive potential, and excessive annual yield (Dumont, Axelsen, et al. 1987). The number of spawning grounds was found to be limited, most of them located in the upstream part of the system. Many previously used sites were no longer accessible or utilized because of various types of human interventions, including dam construction, water pollution, and poaching (Dumont, Axelsen, et al. 1987; La Haye et al. 1992). Three major factors were then proposed to likely account for the high long-term resilience of this stock: the relatively high productivity of the St. Lawrence River; the fact that the intensive commercial fishing was restricted to specific zones, leaving some sectors to act as reservoirs (this "sanctuary" effect is undoubtedly important); and the high selectivity of the historically used commercial gill nets (19 to 20 cm stretched mesh).

Between 1987 and 1991, a new management plan was gradually implemented to reduce the catch, provide more protection to the spawning stock, and strengthen law enforcement. The fishing season and the number of fishing licenses were reduced, longlines, a gear characterized by a low size selectivity favoring the capture of large sturgeon (figure 7) particularly during the colder seasons (Dumont, Axelsen, et al. 1987; Dumont et al. 1989), were banned, and gill net stretch mesh restricted to 20 cm. Sport fishing regulations were also tightened. In the 1990s, research was undertaken to increase knowledge on the characteristics of spawning grounds and juvenile habitats, develop an index of year-class strength in order to anticipate the evolution of the fishery, which harvested mainly fish over 15 to 20 years old, and prevent collapse (La Haye et al. 1992; Nilo, Dumont, and Fortin 1997; Fortin, D'Amours, and Thibodeau 2002). In 1994, the commercial harvest was sampled again in four fishing sectors. In 1997, following a revision of the available new data on the stock and the population, we concluded that the overexploitation status attributed in 1987 was correct and that the population status remained of special concern (Mailhot and Dumont 1998).

Additional data were collected in 1998 to compare the evolution of relative abundance indexes and population dynamic parameters of certain substocks over a 14-year period. This information led to a new diagnosis of the status of this stock and to a revision in the management plan of the fishery. From the beginning of the 1980s to the mid-1990s, growth rate remained about the same. However, for identical lengths, in the four fishing sectors sampled in 1994, sturgeon weights were 7 to 14.8 percent lower than they were a decade earlier. For most 100 mm length classes well represented in the four samples (800 to 1,300 mm), the Fulton condition factor (K) was significantly lower in 1994; the decrease varied between 5 percent and 21 percent and was generally over 10 percent.

Catch curves of the commercial harvest showed that annual mortality rates remained high and, moreover, that the peaks of the catch curves, or the apparent

Figure 7. Lake sturgeon of 78.6 kg caught with a long line in Lac Saint-Louis in March 1975. [Reproduced from Mongeau, Leclerc, and Brisebois 1982.]

age of full recruitment, shifted substantially to older fish in Lac Saint-Louis (from age 16 to 23) and Lac Saint-Pierre (from age 14 to 20). Temporal comparisons of the 20 cm gill net catch per unit of effort showed a reduction of 7 percent in Lac Saint-Louis from 1994 to 1998 and of 25 percent from 1984 to 1998 in the Lac Saint-Pierre archipelago (Mailhot and Dumont 1999). The annual rate of decrease was similar (1.75 percent), but only the second decrease was statistically significant. Juvenile surveys revealed that there has been no rupture in the sequence of cohorts since at least 1980. However, a gradual reduction of 58 percent of the year-class strength index was observed from 1984 to 1992 (Dumont et al. 2000). Two comparatively strong year classes were produced in 1993 and 1994; they were followed by one average and one weak year class. From 1995 to 1999, mark-recapture estimates of the number of mature females on the des Prairies River spawning ground declined by 61 percent, from 1,231 to 500 fish (Fortin, D'Amours, and Thibodeau 2002).

Observed trends in age structure of the harvest (15- to 30-year-old-fish), year-class strength (based on the age distribution of the 2- to 8-year-old segment) and abundance of mature females (over age 25) formed a coherent set of observations indicating an overexploitation status that likely began in the mid-1970s. Combined with the fact that, since 1986, the declared commercial catch remained very high (152–259 metric tons; average, 202 metric tons) and greatly surpassed the historical landings (maximum circa 65 metric tons) reported before 1983 (plate 10), it was clear that the 1987 management plan failed to reverse the decline (Mailhot and Dumont 1999). The observed decrease of the condition factor was seen as an unlikely response to the reduction in abundance of lake sturgeon and possibly related to changes in the trophic conditions of the St. Lawrence River, thus possibly decreasing the potential production of the sturgeon population. A 200-metric-ton commercial catch quota has been enforced since 1999, coupled with the obligation to tag each sturgeon carcass. Individual quotas, based on the fishermen historical catch, were attributed, and severe controls are now applied on fishing sites and all along the commercialization process, including a permit of exportation under the CITES (Convention on International Trade of Endangered Species) agreement. Individual quotas were progressively reduced by 20 percent in 2000, 2001, and 2002 and now total 80 metric tons (or circa 11,500 lake sturgeons). Since 2000, the fishing season has also been shortened by two months and now extends from June 14 to July 31 and September 14 to October 15.

Habitat Conservation and Improvement

The quality and area of five spawning grounds were successfully increased in the des Prairies, Saint-Maurice, Saint-François, Ouareau, and Chaudière rivers (plate 10). In des Prairies River, 8,000 m² of appropriate substrate were added to the most important lake sturgeon spawning area of the system, located downstream from a hydroelectric dam built in 1928 (30 cm of a mix of 20 to 30 cm gravel circled by a 3 m strip of 30 to 50 cm rocks were laid on the plain, unfractured bedrock, under an increasing depth of water from the banks to the middle of the river (maximum 7 m) and exposed to an average current speed of 1 m/s). This project was monitored for three years before and after its realization and provided many useful observations for the management of sturgeon spawning grounds, particularly concerning the decision process to create such habitat, its location relative to water flow and depth, as well as other physical parameters like surface area, substrate, and flow distribution (Fortin, D'Amours, and Thibodeau 2002; Dumont et al. 2011). Egg to larvae survival has been increased from respectively 0.88 and 0.93 percent in 1995 and 1996, before the improvement, to 5.6, 3.82, and 2.41 percent in 1997, 1998, and 1999. Average larval production was also increased from 3.9 million (1.2 to 8.6 million) between 1994 and 1996 to 7 million (2 to 12.8 million) between 1997 and 2003 (Fortin, D'Amours, and Thibodeau 2002; Dumont et al. 2011). As suggested by Russian literature (Khoroshko and Vlasenko 1970) and our own observations, higher survival rates could likely be related to a reduction of egg densities on the spawning beds to under 3,500 eggs/m², a value corresponding, in des Prairies River, to an average spawning bed per female (mean TL, 1300 mm) around 48 m².

During the monitoring period, the highest egg-to-larvae survival rates (1997–1999) were related to an average spawning bed area per female of 13 m². New spawning beds also offered a wider range of suitable conditions of reproduction, and egg and larvae development, under varying annual and seasonal conditions of water discharge (Dumont et al. 2011). When issues like egg overdensity or poor substrate quality are identified, this type of enhancement measure can be used to increase the reproductive success for this species.

In the l'Assomption and Ouareau rivers, two sturgeon spawning beds were exposed to raw sewage, landslides and dam-operated water fluctuation. Governmental and nongovernmental organizations conducted a five-year study (1998–2002) comparing the sturgeon larval production before and after the construction of sewage treatment facilities on the l'Assomption River while monitoring nearby controls in Ouareau River. Massive egg and larval mortality occurred before sewage treatment and a 10-fold improvement in larval production was observed following treatment installation. Study results have raised enthusiasm among authorities and organizations, while the species has become a regional symbol of renewal. The watershed management corporation has recognized the spawning areas as key sites for biodiversity and developed plans for their conservation and restoration (Dumas, Trépanier, and Simoneau 2003).

A major landslide occurred along Ouareau River in March 1990 in the single lake sturgeon spawning ground of this river. Despite the fact that at least two-thirds of the original spawning substrate has been buried and covered by a thick layer of hard clay, and that the flow regime has been deeply modified, the spawning site was still used in the subsequent spring period, and none of the alternative sites contained eggs or larvae (La Haye, Guénette, and Dumont 1990). In winter 2007 and 2008, a total of 18,000 m² of spawning bed was added near this existing spawning site by using appropriate substrate and by building two low-height rock dams in order to prevent eggs and fry from drying up in a river submitted to extreme variations of water flow. Sturgeon egg deposition was observed in 2007 and 2008 on both artificial and natural sites.

In 1999, a new spawning area (1,300 m²) for lake sturgeon was created downstream of La Gabelle dam on Saint-Maurice River to compensate for an encroachment on fish habitat related to the refurbishment of the power dam by Hydro Québec. Monitoring in 2000 and 2001 showed that the spawning bed was used under a wide range of flow conditions by lake sturgeon and at least four other species (GDG Conseil Inc. 2001). Similar projects were also realized in the Saint-François (La Haye, Clermont, and Lemire 1996) and Chaudière (Trencia and Collin 2006) rivers in order to improve spawning bed quality and area in sectors historically used by lake sturgeon.

Finally, another spawning ground was artificially created downstream from Beauharnois power dam, near a site of concentration of large mature sturgeons, but this trial has not been successful. Substrate silting, thick growth of aquatic vegetation, and frequent flow changes near the spawning area were identified as the main factors limiting the attraction for spawners (Gendron, Lafrance, and La Haye 2002). In the same sector, during the 1980s, new operating and discharge criteria were tested and applied to restore access to a spawning ground located in the residual bed of the St. Lawrence River at its confluence in Lac Saint-Louis, and historically used before the construction of the Pointe-des-Cascades dam in 1964.

Dredging projects along the navigation channels and in harbor works are commonly analyzed and modified in order to protect adult and juvenile habitats. The construction of fishways might help alleviate the problem of habitat fragmentation, but would not remedy the loss of habitat resulting from the modification of flow regime. The swimming ability of sturgeon is different from that provided for in current fishway designs, and their large size further complicates design of adequate passage (see Peake et al. 1997).

Each spring since 2001, only a few large lake sturgeon (<40) are among the thousands of fish, belonging to at least 36 species, that use the Vianney-Legendre fishway built at the Saint-Ours dam, on the Richelieu River (Fleury and Desrochers 2004). As reported by Paradis and Malo (2003), this single vertically slit ladder, composed of 16 successive large basins (with a 15 cm level difference between each basin), was designed to respond to the requirements and characteristics of lake sturgeon, copper redhorse, river redhorse (*Moxostoma carinatum*), American shad (*Alosa sapidissima*), and American eel (*Anguilla rostrata*). Data from a recent study on lake sturgeon behavior and passage success in this fishway contribute to understanding how these structures can be used to facilitate the upstream passage of Acipenserids at dams (Thiem et al. 2011).

A study was initiated in 2001 to quantify the impacts of water discharge variation on the St. Lawrence River fish community. Habitat modeling emerged from a creative multidisciplinary collaboration with modelers, integrating the physical dimension of the habitat with biological processes in a 2-D spatially explicit model. The digital model covered a large part of the lake sturgeon distribution area in the

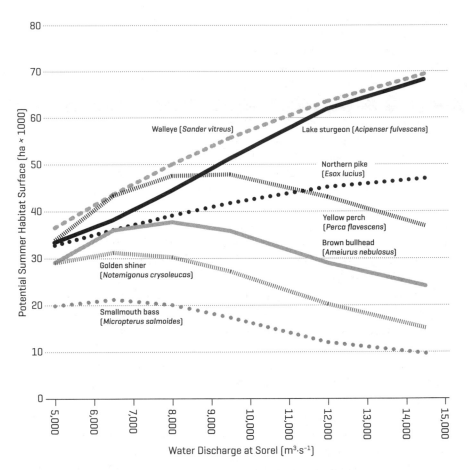

Figure 8. Relation between the St. Lawrence River discharge and area of suitable habitat for various species representative of the fish community. From the bottom up: *Micropterus salmoides, Notemigonus crysoleucas, Ameiurus nebulosus, Perca flavescens, Esox lucius, Acipenser fulvescens,* and *Sander vitreus.* [Adapted from Mingelbier et al. 2004.]

Quebec section of the St. Lawrence River with a high spatial resolution. Surfaces of suitable lake sturgeon habitat, estimated for a broad spectrum of hydrological conditions, were found to increase with discharge (figure 8). Recommendations concerning fish habitat protection in the fluvial St. Lawrence were presented to the International Joint Commission under the "in process" evaluation of the regulation criteria of the Lake Ontario–St. Lawrence River system (Mingelbier, Brodeur, and Morin 2004, 2005a, 2005b).

In 2006, a hydroelectric project in the Courant Sainte-Marie, an important rapid in front of Montreal, was withdrawn to protect one of the two remaining rapid sectors of the St. Lawrence River and to maintain free movements of such migratory species as lake sturgeon, American shad, American eel, and copper redhorse (Dumont, Bilodeau, and Leclerc 2005). In the mid-1980s, a similar project in the Lachine rapids about 15 km upstream was also rejected for similar environmental concerns.

The Future

A first evaluation of the combined effects on recruitment of habitat improvement efforts in the 1990s and of the 2000–2002 management plan was realized in 2007 and suggests that these measures likely contributed to a significant increase of lake sturgeon recruitment in the lower St. Lawrence River lake sturgeon population. Recruitment has been continuous at least since 1980, but has also been highly variable (figure 9A). The strength of year classes 1984 to 2004 appeared mainly influenced by hydrological factors, strong year-classes being associated with des Prairies River June water discharges over 11,150m³/s (Nilo, Dumont, and Fortin; Dumont et al. 2011), confirming the great importance of this river spawning grounds for the lower St. Lawrence River lake sturgeon population (figure 9B). However, larval production estimates from 1994 to 2003 also indicate that strong cohorts, like those of 1994 and 2002, were related to high larval drift in this river (figure 9C). Larval production appeared independent from spring river discharge ($r = 0.07$; $p < 0.05$), but was strongly and inversely related to the previous year commercial landings in the St. Lawrence River (figure 9D), suggesting that annual quota reductions applied in 2000, 2001, and 2002 led to an annual 35 percent increase of larval drift in the des Prairies River.

Lake sturgeon are sensitive to habitat degradation and fragmentation, and to overfishing. After a long period of decline in the lower St. Lawrence River, large and increasing production of larvae in the des Prairies River, strong year-class appearance, increasing abundance of subadult fishes in Lac Saint-Louis (Mailhot, Dumont, and Vachon 2011), the recent return in abundance of lake sturgeon on disused spawning grounds in the Chaudière and Richelieu rivers (MRNF, unpublished data), and positive comments from commercial fishermen are recent encouraging signs of improvement of the status of the lake sturgeon population even if, as recognized by most experienced commercial fishermen, the current abundance is clearly lower than it was 30 years ago.

Recent improvement is likely partly related to our sustained effort of management of this unique population during the past 25 years. In the future, because of the complexity of the system and the simultaneous influence of many factors on the

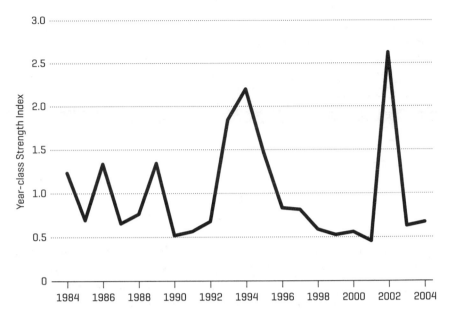

Figure 9a. Variation of the lake sturgeon year-class strength index in the lower St. Lawrence River, 1984–2004.

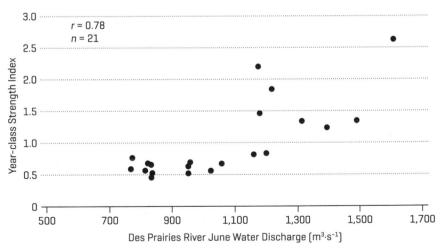

Figure 9b. During this period strong year-classes were related to mean June water discharges over 1,150 m³/s Des Prairies River.

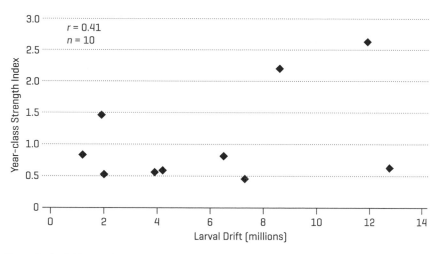

Figure 9c. Available data for the period 1994 to 2003 indicate that there was no correlation between year-class strength and larval production [$r = 0.41$; $n = 10$; $p > 0.1$], but the two largest cohorts, those in 1994 and 2002, were associated with high larval production. On the other hand, the smaller-than-average 2003 cohort was associated with the highest larval production measured during the period.

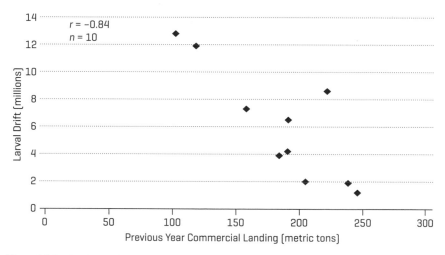

Figure 9d. During this period, larvae production appeared highly and negatively related to the commercial landings in the previous year. In the 1990s, larval production decreased as the previous year's recorded landings increased, and it rose starting in the year 2000 as the previous year's landings began to decrease following the gradual application of a progressively lower commercial annual harvest quota [from 200 t in 1999 to 80 t in 2002]. These observations suggest that larvae production quickly and positively responded to the implement of a decreasing quota between 2000 and 2002 but also demonstrate that environmental factors can alter the makeup of a cohort at a later stage. [Adapted from Dumont et al. 2011 and Mailhot et al. 2011.]

lower St. Lawrence River sturgeon population, its habitat and exploitation, it will be important to achieve the following protective measures:

- Prevent additional fragmentation of this 350 km stretch of fluvial habitat
- Maintain the application to the fishery of conservative restrictions, measures of control, law enforcement, and periodic monitoring
- Preserve and improve the quality of the known spawning grounds, feeding habitats, and deep water refuges, whose quality and areas will be reduced along with the St. Lawrence River flow reduction predicted by the climate changes models
- Intensify the efforts of reduction of water pollution in the Great Lakes–St. Lawrence River system
- Continue to deepen our knowledge of the biology and habitat of this species.

NOTE

This chapter is dedicated to Réjean Fortin, who died prematurely in 2001. Réjean was an excellent scientist and pedagogue, and a great colleague. His contribution to the knowledge of sturgeon biology will remain pertinent for future decades. We also want to underline the work of J-P. Cuerrier, G. Roussow, V. D. Vladykov, E. Magnin, and J.-R. Mongeau, which still continues to inspire our collective effort to develop a comprehensive approach to the conservation of the St. Lawrence River lake sturgeon population. Finally, we wish to acknowledge the following persons for their involvement over the last 25 years in this long-term objective : L. Aubry, R. Bacon, M. Bernard, P. Bilodeau, A. Blanchard, A. Branchaud, V. Boivin, D. Bourbeau, L. Bouthillier, J. Brisebois, P. Brodeur, Y. Chagnon, S. Clermont, P. Y. Collin, C. Côté, J. D'Amours, M. Damphousse, G. Desjardins, S. Desloges, D. Dolan, B. Dumas, R. Dumas, R. Faucher, D. Fournier, N. Fournier, S. Garceau, D. Goyette, S. Guénette, F. Guilbard, D. Hatin, M. Lafleur, M. La Haye, J. Leclerc, P. Leclerc, M. Léveillé, G. Massé, H. Massé, M. Mingelbier, J. Morin, P. Nilo, G. Ouellet, A. Paquet, Y. Poiré, J. Robitaille, G. Roy, M. Rousseau, S. Thibodeau, G. Trencia, F. Veillette, and R. Verdon. M. Courtemanche provided interesting archaeological information on lake sturgeon use by First Nations people.

REFERENCES

Beamish, F.W.H., D.L.G. Noakes, and A. Rossiter. 1998. Feeding ecology of juvenile Lake Sturgeon, *Acipenser fulvescens*, in northern Ontario. Canadian Field-Naturalist 112:459-468.

Boucher, P. 1664. Histoire véritable et naturelle des mœurs et productions du pays de la Nouvelle-France vulgairement dite le Canada. Société historique de Boucherville, Québec, 1964.

Callaway, C. G. 1989. Indians of North America: The Abénaki. Chelsea House.

Clermont, N., C. Chapdelaine, and J. Cinq-Mars. 2003. L'Ile aux Alumettes. L'Archaïque supérieur dans l'Outaouais. Paléo-Québec 30, Recherches Amérindiennes au Québec,

Montréal.

Courtemanche, M. 2003. Pratiques halieutiques à la station 4 de la Pointe-du-Buisson (BhF1-1) au Sylvicole Moyen Tardif (920–940 A.D.). M. Sc. thesis, Département d'Anthropologie, Université de Montréal, Québec.

Cuerrier, J-P. 1945. Les stades de maturité chez l'esturgeon du lac Saint-Pierre. Office de Biologie, MS 2212, Ministère de la Chasse et des Pêcheries, Québec.

———. 1962. Aperçu général sur l'inventaire biologique des poissons et des pêcheries de la région du lac Saint-Pierre. Naturaliste canadien 89:193–213.

———. 1966. Observations sur l'esturgeon de lac *Acipenser fulvescens* Raf. dans la région du lac Saint-Pierre au cours de la période de frai. Naturaliste canadien 93:279–334.

Cuerrier, J.-P., F. E. J. Fry, and G. Préfontaine. 1946. Liste préliminaire des poissons de la région de Montréal et du lac Saint-Pierre. Naturaliste canadien 73:17–32.

Cuerrier, J.-P., and G. Roussow. 1951. Age and growth of lake sturgeon from Lake St. Francis, St. Lawrence River. Canadian Fish Culturist 10:17–29.

D'Amours, J., S. Thibodeau, and R. Fortin. 2001. Comparison of lake sturgeon (*Acipenser fulvescens*), *Stizostedion* spp, *Catostomus* spp, *Moxostoma* spp., quillback (*Carpiodes cyprinus*) and mooneye (*Hiodon tergisius*) larval drift in des Prairies River, Québec. Canadian Journal of Zoology 79:1472–1489.

Doyon, C., S. Boileau, R. Fortin, and P. A. Spear. 1998. Rapid HPLC analysis of retinoids and dehydroretinoids stored in fish liver: Comparison of two lake sturgeon populations. Journal of Fish Biology 53:973–976.

Doyon, C., R. Fortin, and P. A. Spear. 1999. Retinoic acid hydroxylation and teratogenesis in lake sturgeon (*Acipenser fulvescens*) from the St. Lawrence River and Abitibi region, Québec. Canadian Journal of Fisheries and Aquatic Science 56: 1428–1436.

Dubreuil, R., and J.-P. Cuerrier. 1950. Cycle de maturation des glandes génitales chez l'esturgeon de lac (*Acipenser fulvescens*, Raf.). Institut de Biologie générale et de Zoologie, Université de Montréal, Québec.

Dumas, R., F. Trépanier, and M. Simoneau. 2003. Fish problems and partnership solutions: The lake sturgeon case in the L'Asomption watershed. American Fisheries Society 133rd Annual Meeting, Québec City, Canada, August 10–14.

Dumont, P., F. Axelsen, H. Fournier, P. Lamoureux, Y. Mailhot, C. Pomerleau, and B. Portelance. 1987. Avis scientifique sur le statut de la population d'Esturgeon jaune dans le système du fleuve Saint-Laurent. Ministère du Loisir, de la Chasse et de la Pêche du Québec et ministère de l'Agriculture, des Pêcheries et de l'Alimentation du Québec. Comité scientifique conjoint. Avis scientifique 87/1. 21 p.

Dumont, P., P. Bilodeau, and J. Leclerc. 2005. Portrait sommaire de la faune ichtyologique du Courant Sainte-Marie (fleuve Saint-Laurent). Travail réalisé pour le Comité du Bassin du Havre, Ministère des Ressources naturelles et de la Faune, Longueuil, Québec.

Dumont, P., J. D'Amours, S. Thibodeau, N. Dubuc, R. Verdon, S. Garceau, P. Bilodeau, Y. Mailhot, and R. Fortin. 2011. Effects of the development of a newly created spawning ground in the des prairies River (Quebec, Canada) on the reproductive success of lake sturgeon *Acipenser fulvescens*. Journal of Applied Ichtyology 27:394-404.

Dumont, P., R. Fortin, G. Desjardins, and M. Bernard. 1987. Biology and exploitation of lake sturgeon (*Acipenser fulvescens*) in the Québec waters of the Saint-Laurent River. *In*: Proceedings of a workshop on lake sturgeon (*Acipenser fulvescens*), Feb. 27–28 1986, Timmins. C. H. Olver, ed. Ontario Fisheries Technical Report. Series 23: 57–76.

Dumont, P., P. Lamoureux, G. Laforce, M. La Haye, and N. Fournier. 1989. Influence de la dimension de l'hameçon sur la sélectivité et le rendement de la ligne dormante pour la capture de l'esturgeon jaune (*Acipenser fulvescens*). Québec, Ministère du Loisir, de la Chasse et de la Pêche et Ministère de l'Agriculture, des Pêcheries et de l'Alimentation, Avis scientifique 89/1.

Dumont, P., J. Leclerc, J.-D. Allard, and S. Paradis. 1997. Libre passage des poissons au barrage de Saint-Ours, rivière Richelieu. Québec, ministère de l'Environnement et de la Faune, Direction régionale de la Montérégie et Direction des ressources matérielles et des immobilisations, et ministère du Patrimoine canadien (Parcs Canada).

Dumont, P., J. Leclerc, Y. Mailhot, E. Rochard, C. Lemire, H. Massé, Hélène Gouin, Denis Bourbeau, and Daniel Dolan. 2000a. Suivi périodique de l'évolution du recrutement de l'esturgeon jaune en 1999. *In*: Compte rendu du cinquième atelier sur les pêches commerciales, Duchesnay, 18–20 janvier 2000. M. Bernard and C. Groleau, eds. Québec, Société de la faune et des parcs du Québec.

Dumont, P., Y. Mailhot, R. Dumas, and P. Bilodeau. 2000b. Plan de gestion de l'esturgeon jaune du fleuve Saint-Laurent 2000–2002. Société de la faune et des parcs du Québec, Directions de l'aménagement de la faune du Centre-du-Québec, de Lanaudière, de la Montérégie et de Montréal .

Fleury, C., and D. Desrochers. 2004. Validation de l'efficacité des passes à poisson au lieu historique national du Canal-de-Saint-Ours saison 2003. Rapport final préparé pour Parcs Canada par Milieu Inc, Laprairie, Québec.

Fortin, R., J. D'Amours, and S. Thibodeau. 2002. Effets de l'aménagement d'un nouveau secteur de frayère sur l'utilisation du milieu en période de fraie et sur le succès de reproduction de l'esturgeon jaune (*Acipenser fulvescens*) à la frayère de la rivière des Prairies. Rapport synthèse 1995–1999. Pour l'Unité Hydraulique et Environnement, Hydro-Québec et la Société de la faune et des parcs du Québec, Direction de l'aménagement de la faune de Montréal, de Laval et de la Montérégie. Département des Sciences biologiques, Université du Québec à Montréal.

Fortin, R., P. Dumont, and S. Guénette. 1996. Determinants of growth and body condition of lake sturgeon (*Acipenser fulvescens*). Canadian Journal of Fisheries and Aquatic Science 53:1150–1156.

Fortin, R., S. Guénette, and P. Dumont. 1992. Biologie, modélisation et gestion des populations d'esturgeon jaune (*Acipenser fulvescens*) dans 14 réseaux de lacs et de rivières du Québec. Québec, Ministère du Loisir, de la Chasse et de la Pêche, Direction régionale de Montréal et Service de la faune aquatique, Montréal et Québec.

Fortin, R., J.-R. Mongeau, G. Desjardins, and P. Dumont. 1993. Movements and biological statistics of lake sturgeon (*Acipenser fulvescens*) populations from the St. Lawrence and Ottawa River system, Québec. Canadian Journal of Zoology 71:638–650.

Fry, F. E. J., G. Préfontaine, et al. 1941. Alimentation de quelques espèces de poisson du lac Saint-Louis et du lac des Deux-Montagnes. Rapport de la Station biologique de Montréal, pour l'année 1941. Ministère de la Chasse et des Pêcheries, Québec, Fascicule II, Appendice X, 188–218.

Garceau, S., and P. Bilodeau. 2004. La dérive larvaire de l'esturgeon jaune (*Acipenser fulvescens*) à la rivière des Prairies, aux printemps 2002 et 2003. Ministère des Ressources naturelles, de la Faune et des Parcs, Direction de l'aménagement de la faune de Montréal, de Laval et de la Montérégie, Longueuil, Rapport technique 16-21.

GDG Conseil Inc. 2001. Réfection de la centrale de La Gabelle. Programme de surveillance et de suivi environnemental. Utilisation par les poissons d'un nouveau secteur de fraie aménagé en aval de la centrale de la Gabelle—printemps 2001. Rapport présenté à Hydro-Québec, vice-présidence Exploitation des équipements de production. Unité Hydraulique et Environnement.

Gendron, M., P. Lafrance, and M. La Haye. 2002. Suivi de la frayère en aval de la centrale Beauharnois. Environnement Illimité inc. pour Hydro Québec, Division Production, Montréal.

Giguère, G.-E. 1973. Oeuvres de Champlain. Éditions du Jour.

Guénette, S., R. Fortin, and E. Rassart. 1993. Mitochondrial DNA variation in lake sturgeon (*Acipenser fulvescens*) from the St. Lawrence River and James Bay drainage basins in Québec, Canada. Canadian Journal of Fisheries and Aquatic Science 50:659–664.

Guénette, S., D. Goyette, R. Fortin, J. Leclerc, N. Fournier, G. Roy, and P. Dumont. 1992. La périodicité de la croissance chez la femelle de l'esturgeon jaune (*Acipenser fulvescens*) du fleuve Saint-Laurent est-elle reliée à la périodicité de la reproduction? Canadian Journal of Fisheries and Aquatic Science 49:1336–1342.

Guénette, S., E. Rassart, and R. Fortin. 1992. Morphological differentiation of lake sturgeon (*Acipenser fulvescens*) from the St. Lawrence River and Lac des Deux Montagnes (Québec, Canada). Canadian Journal of Fisheries and Aquatic Science 49:1959–1965.

Guilbard, F., J. Munro, P. Dumont, D. Hatin, and R. Fortin. 2007. Feeding ecology of Atlantic and lake sturgeon in the St. Lawrence estuarine transition zone. American Fisheries Society Symposium 56:85–104.

Harkness, W. J. K., and J. R. Dymond. 1961. The lake sturgeon: The history of its fishery and problems of conservation. Ontario Department of Lands and Forest, Fish and Wildlife Branch.

Hatin, D., S. Lachance, and D. Fournier. 2007. Effect of dredged sediment deposition on use by Atlantic sturgeon and lake sturgeon at an open-water disposal site in the St. Lawrence estuarine transition zone. American Fisheries Society Symposium 56:235–255.

Haxton, T. J. 2008. A synoptic review of the history and our knowledge of lake sturgeon in the Ottawa River. Ontario Ministry of Natural Resources, Southern Science and Information Technical Report SSI 126.

Haxton, T. J., and C. S. Findlay. 2008. Variation in lake sturgeon abundance and growth among river reaches in a large regulated river. Canadian Journal of Fisheries and Aquatic Sciences 65:646–657.

Joliff, T. W., and T. H. Eckert. 1971. Evaluation of present and potential sturgeon fisheries of the St. Lawrence River and adjacent waters. New York Department of Environmental Conservation, Cape Vincent Fisheries Station.

Khoroshko, P. N., and A. D. Vlasenko. 1970. Artificial spawning grounds of sturgeon. Journal of Ichthyology 10:286–292.

La Haye, M., A. Branchaud, M. Gendron, R. Verdon, and R. Fortin. 1992. Reproduction, early life history, and characteristics of the spawning grounds of the lake sturgeon (*Acipenser fulvescens*) in des Prairies and L'Assomption rivers, near Montreal, Québec. Canadian Journal of Zoology 70:1681–1689.

La Haye, M., S. Clermont, and C. Lemire. 1996. Localisation d'une frayère à esturgeons jaunes dans le cours inférieur de la rivière St-François. Enviro-science inc pour l'Association des Pêcheurs commerciaux du lac Saint-Pierre, Nicolet, Québec.

La Haye, M., S. Desloges, C. Côté, J. Deer, S. Philips Jr., B. Giroux, S. Clermont, and P. Dumont.

2003. Location of lake sturgeon (*Acipenser fulvescens*) spawning grounds in the upper part of the Lachine rapids. Société de la faune et des parcs du Québec, Direction de l'aménagement de la faune de Montréal, de Laval et de la Montérégie, Longueuil, Technical Report 16-15E.

La Haye, M., S. Desloges, C. Côté, A. Rice, S. Philips Jr., J. Deer, B. Giroux, K. de Clerk, and P. Dumont. 2004. Search for and characterization of lake sturgeon (*Acipenser fulvescens*) spawning grounds in the upstream portion of the Lachine Rapids, St. Lawrence River, in 2003. Ministère des Ressources naturelles, de la Faune et des Parcs, Direction de l'aménagement de la faune de Montréal, de Laval et de la Montérégie, Longueuil, Technical Report 16-20E.

La Haye M., S. Guénette, and P. Dumont. 1990. Utilisation de la frayère de la rivière Ouareau par l'Esturgeon jaune suite à l'éboulis survenu en mars 1990. Québec, Ministère du Loisir, de la Chasse et de la Pêche, Direction régionale de Montréal. Rapport technique 06-07.

Magnin, E. 1962. Recherches sur la systématique et la biologie des Acipenséridés *Acipenser sturio* L., *Acipenser oxyrhynchus* Mitchill et *Acipenser fulvescens* Raf. Thèse présentée à la faculté des sciences de l'Université de Paris. Imprimerie nationale.

———. 1977. Croissance, régime alimentaire et fécondité des esturgeons *Acipenser fulvescens* Rafinesque du bassin hydrographique de la Grande rivière (Québec). Naturaliste canadien 104:419–427.

Magnin, E., and G. Beaulieu. 1960. Déplacements des esturgeons *Acipenser fulvescens* et *A. oxyrhyncus* du fleuve Saint-Laurent d'après les données de marquage. Naturaliste canadien 87:237–252.

Mailhot, Y., and P. Dumont. 1998. Avis scientifique: Révision du statut du stock d'esturgeon jaune du fleuve Saint-Laurent. *In*: Compte rendu du troisième atelier sur les pêches commerciales, Duchesnay, Ministère de l'Environnement et de la faune du Québec, Québec, 13–15 janvier 1998. M. Bernard and C. Groleau, eds.

———. 1999. Mise à jour de l'état du stock d'esturgeon jaune du fleuve Saint-Laurent. *In*: Compte rendu du quatrième atelier sur les pêches commerciales, Duchesnay, Faune et Parcs Québec, Québec, 12–14 janvier 1999. M. Bernard and C. Groleau, eds.

Mailhot, Y., P. Dumont, and N. Vachon. 2011. Management of the lake sturgeon *Acipenser fulvescens* population in the lower St. Lawrence River (Québec, Canada) from the 1910s to the present. Journal of Applied Ichtyology 27: 405–410.

Mélançon, C. 1936. Les poissons de nos eaux. Librairie Granger.

Mingelbier, M., P. Brodeur, and J. Morin. 2004. Impacts de la régularisation du débit des Grands Lacs et des changementss climatiques sur l'habitat du poisson du fleuve Saint-Laurent. Vecteur Environnement 37(6): 34–43.

———. 2005a. Recommendations concerning fish and their habitats in the fluvial St. Lawrence and evaluation of the regulation criteria for the Lake Ontario–St. Lawrence River system. Report presented to the International Joint Commission. Ministère des Ressources naturelles et de la Faune du Québec, Direction de la recherche faunique, Québec.

———. 2005b. Modélisation numérique 2D de l'habitat des poissons du Saint-Laurent fluvial pour évaluer l'impact des changements climatiques et de la régularisation. Le Naturaliste Canadien 129:96–102.

Mongeau, J.-R., J. Leclerc, and J. Brisebois. 1982. La dynamique de la reconstitution des populations de l'esturgeon jaune, *Acipenser fulvescens*, du lac des Deux-Montagnes, province de Québec, de 1964 à 1979. Québec, Ministère du Loisir, de la Chasse et de la

Pêche, Rapport technique 06-33.

Montpetit, C. 1897. Les poissons d'eau douce du Canada. C.-O. Beauchemin & Fils.

Morin, J., and M. Leclerc. 1998. From pristine to present state: Hydrology evolution of Lake Saint-François, St. Lawrence River. Canadian Journal of Civil Engineering 25: 864–879.

Ndayibagira, A., M.-J. Cloutier, P. D. Anderson, and P. A. Spear. 1995. Effects of 3,3,' 4.4'-tetrachlorobiphenyl on the dynamics of the vitamin A in brook trout (*Salvelinus fontinalis*) and intestinal retinoid concentration in lake sturgeon (*Acipenser fulvescens*). Canadian Journal of Fisheries and Aquatic Science 52:512–520.

Nilo, P., P. Dumont, and R. Fortin. 1997. Climatic and hydrological determinants of year-class strength of St. Lawrence River lake sturgeon (*Acipenser fulvescens*). Canadian Journal of Fisheries and Aquatic Science 54:774–780.

Nilo, P., S. Tremblay, A. Bolon, J. Dodson, P. Dumont, and R. Fortin. 2007. Feeding ecology of juvenile lake sturgeon *Acipenser fulvescens* in the St. Lawrence River system. Transactions of the American Fisheries Society 135:1044–1055.

Paradis, S., and R. Malo. 2003. Efficiency of the Vianney-Legendre fish ladders at the Saint-Ours Canal National Historical Site, Richelieu River, Quebec. American Fisheries Society 133rd Annual Meeting, Québec City, Canada, August 10–14.

Peake, S., F. W. H. Beamish, R. S. McKinley, D. A. Scruton, and C. Katopodis. 1997. Relating swimming performance of lake sturgeon, *Acipenser fulvescens*, to fishway design. Canadian Journal of Fisheries and Aquatic Sciences 54:1361–1366.

Robitaille, J. A., Y. Vigneault, G. Shooner, C. Pomerleau, and Y. Mailhot. 1988. Modifications physiques de l'habitat du poisson dans le Saint-Laurent de 1945 à 1984 et effets sur les pêches commerciales. Canadian Technical Report of Fisheries and Aquatic Sciences 1808.

Rousseaux, C. G., A. Branchaud, and P. A. Spear. 1995. Evaluation of liver histopathology and erod activity in St. Lawrence lake sturgeon (*Acipenser fulvescens*) in comparison with a reference population. Environmental Toxicology and Chemistry 14:843–849.

Roussow, G. 1955a. Quelques observations sur les variations de forme et de couleur chez les esturgeons de la province de Québec. Annales de l'Acfas 21:79–85.

———. 1955b. Les esturgeons du fleuve Saint-Laurent en comparaison avec les autres espèces d'Acipenséridés. Office de Biologie, Ministère de la Chasse et des Pêcheries, Province de Québec, Montréal.

———. 1957. Some considerations concerning sturgeon spawning periodicity. Journal of Fisheries Research Board of Canada 14:553–572.

Scott, W. B., and E. J. Crossman. 1973. Freshwater fishes of Canada. Fisheries Research Board of Canada. Bulletin 184.

Thiem J.D., T. R. Binder, J. W. Dawson, P. Dumont, D. Hatin, C. Katopodis, D. Z. Zhu, and S. J. Cooke. 2011. Behaviour and passage success of upriver-migrating lake sturgeon *Acipenser fulvescens* in a vertical slot fishway on the Richelieu River, Quebec, Canada. Endangered Species Research 15:1–11.

Trencia, G., and P.-Y. Collin. 2006. Rapport d'aménagement d'une frayère pour le poisson à la rivière Chaudière. Ministère des Ressources naturelles et de la Faune du Québec, Direction de l'Aménagement de la Faune Chaudière-Appalaches, Lévis.

Veillette, F. 2007. Étude de différents indicateurs biologiques chez l'esturgeon jaune (*Acipenser fulvescens*) du Québec. M. Sc. Thesis. Université du Québec à Montréal.

Vladykov, V. D. 1948. Rapport du biologiste du département des pêcheries. 2. Nourriture de l'esturgeon. *In*: Rapport général du Ministre de la Chasse et des Pêcheries de la Province de Québec concernant les activités du département des pêcheries pour l'exercice financier 1947–48, Québec.

———. 1955. Fishes of Quebec. Sturgeons. Album 5. Department of Fisheries, Quebec.

Vladykov, V. D., and G. Beaulieu. 1946. Études sur l'esturgeon (*Acipenser*) de la province de Québec. I: Distinction entre deux espèces d'esturgeons par le nombre de boucliers osseux et de branchiospines. Naturaliste canadien 73:143–204.

———. 1951. Études sur l'esturgeon (*Acipenser*) de la province de Québec. II: Variations du nombre de branchiospines sur le premier arc branchial. Naturaliste canadien 78:129–154.

Vladykov, V. D., and C. Gauthier. 1941. Remarques sur le régime alimentaire de l'esturgeon (*A. fulvescens*) dans le lac Saint-Louis. Rapport de la Station biologique de Montréal, pour l'année 1941. Ministère de la Chasse et des Pêcheries, Québec, Fascicule III, Appendice IX: 384–387.

Vladykov, V. D., and J. R. Greely. 1963. Order Acipenseroidei. *In*: Fishes of the Western North Atlantic, no. 1, part 3. Soft-rayed bony fishes. H. B. Bigelow, ed. Memoir Sears Foundation of Marine Research, Yale University.

Welsh, A., T. Hill, H. Quinlan, C. Robinson, and B. May. 2008. Genetic assessment of lake sturgeon population structure in the Laurentian Great Lakes. North American Journal of Fisheries Management 28:572–591.

MARTY HOLTGREN

Bringing Us Back to the River

The annual nmé (lake sturgeon) return and its celebration by our Peoples assure the renewal and continuation of human and all other life.

— Little River Band of Ottawa Indians Nmé Stewardship Plan

THE HISTORY OF *NMÉ* IN THE GREAT LAKES IS A STORY OF HARMONY, TRAGEDY, and an opportunity for redemption. The tragedy has been well documented by historians, academics, and writers chronicling a time when sturgeon were an abundant member of the lakes and their later spiral toward near extinction (Auer 1999; Tody 1974; Schoolcraft 1970; Harkness and Dymond 1961). I have sat through many lectures on sturgeon where I was presented with the same exhausting information about how their habitat was destroyed, how they were overharvested, and why the current outlook is so bleak. I believe it is time for a different story for the sturgeon, one of harmony and commitment toward their recovery, where the story comes from many voices.

One such source is from Native American people who have lived in harmony with sturgeon for millennia. Their relationship with the sturgeon may be characterized by conservation approaches, stewardship, and religious beliefs (LRBOI 2008; Rettig, Berkes, and Pinkerton 1989). Native American oral history provides us information about a time when the sturgeon and human communities were in balance within the Great Lakes ecosystem. This history, and the worldview that comes with it, can provide a road map to restoring this species' place in the Great Lakes—an opportunity for redeeming the tragedy that has been so well documented.

This opportunity for the Little River Band of Ottawa Indians (LRBOI) came in the form of a cargo trailer on the banks of the Big Manistee River.

When I tell people that I am a fisheries biologist for the LRBOI and doing sturgeon restoration, I usually get a blank look followed by puzzlement and then a set of common questions. "Why would a tribe have a natural resources department and be involved with *managing* fish?" "Would a tribal *management* approach be different from that of the other agencies?" Within this chapter I hope to make apparent the answer to these questions by detailing the road map to redemption I spoke of earlier. I will use a case study from the LRBOI to demonstrate a unique tribal approach to restoration and stewardship.

The answer to the first question is embedded in a deep history of cultural, social, and political elements. Simply put, the tribes in the Great Lakes manage natural resources to protect cultural sovereignty and to meet the generational and unique needs of tribal members. Cultural sovereignty is the process of tribes making decisions internally that protect traditions and customs (Coffey and Tsosie 2001), and this is evident throughout the Great Lakes, as the tribes are managing watersheds, reservation natural resources, and species that are of great consequence to them. The tribes have a *need* to manage the fishery in addition to other management institutions because tribal needs and worldviews are often very different from those of the general population (Berkes 2009; Mattes and Kmiecik 2006; Kimmerer 2000; Salmon 2000; Notske 1995; Busiahn 1989).

The sturgeon population in the Big Manistee River, Michigan, is an example where for over 100 years the population was overlooked, where decisions and concerns for resource managers were often how many exotic trout should be stocked or what the fisherman's opinion may be about that year's harvest. Out of necessity and a cultural responsibility, the tribes began to work on restoring the sturgeon because it was unacceptable to them to lose a species that is revered and belongs in the watershed (LRBOI 2008).

So why different management strategies for different agencies? The basis of state fisheries management relies on ownership of the fisheries resource. This is known as the common property principle, where the entire populace owns the fishery and the state government has the right and responsibility of being the trustee (Henquinet and Dobson 2006; Nielsen 1999). Within this framework, the state has the difficult task of maintaining open access to the fishery while ensuring the protection, sustainability, and productivity of the resource. Over the past century many of the species that are important to the tribes did not receive priority under the auspices of the public trust. Examples are abundant; in the state of Michigan millions of dollars have been spent on introducing nonnative species with little funding being spent on endemic species of significance to the tribes, including lake sturgeon, sucker, and the extirpated Arctic grayling. Therefore, the tribes have a *need* to keep these species present regardless of what the current state management perspective or funding priorities may be.

The LRBOI has a past that recognizes the importance of sturgeon and a future that is focused on preserving it (LRBOI 2008; McClurken 2009). The Big Manistee River defines the tribe's reservation, which is one of the few rivers on the eastern shoreline of Lake Michigan supporting a sturgeon population known to have a small group of spawners. Historically, tribal people would gather on the banks of the river each year for the lake sturgeon, sucker, and Arctic grayling spawning runs. Jay Sam, tribal cultural preservation director, says this about this historic event: "The grandfather fish (sturgeon), and its relatives the undermouth fish (sucker), they would sacrifice themselves during the sucker moon so the people would have food until the other crops were available." The sturgeon is a clan spirit (LRBOI 2008).

The historical importance of these clan spirits within tribal communities is evidenced on the pages of documents from the 1800s, where Tribal people would "sign their names" not in English but with symbols and their Aniishinabek names that represented their family lineage and clans (LRBOI 2008; GLIFWC 2007). On these documents is the distinct image of the sturgeon. These clan spirits are often the focus of tribal natural resources departments: the LRBOI, Menominee Indian Tribe of Wisconsin (Runstrom et al. 2002), and White Earth Nation with sturgeon, the Little Traverse Bay Bands of Odawa Indians with wolves, and the Grand Traverse Band of Ottawa and Chippewa Indians with martin are just a few examples.

In 1836 a treaty was signed where five tribes ceded about one-third of what is now known as the state of Michigan. The cession guaranteed that the tribes would be granted the "Usual privileges of occupancy" to hunt, fish, and gather from the land and waters. I cannot help but wonder how the tribal chiefs at the time of the treaty signing viewed the Great Lakes. Could they envision a time when the clan spirits and the species so vital to their communities would be gone? The list is exhaustive. The Arctic grayling and woodland caribou have gone extinct, and the populations of wolf, martin, and moose are only a fraction of what they once were. This is where the tribes have continued to play a large role in management by participating in natural resource management decision making, developing tribal stewardship plans, conducting biological assessments and restoration projects.

New Beginnings

In 2001 the LRBOI began to quantify fish populations within its reservation waters. Past research had shown that the sturgeon population was very small, and no true evidence of recent natural reproduction was available. The tribe hired fisheries professionals, purchased equipment, and began collecting information on the lake sturgeon. From the very beginning the tribal community was excited about the sturgeon program and at the prospect of seeing them restored. The excitement produced what was called a "Cultural Context Group" made up of tribal members and biologists that would develop goals and objectives for their sturgeon program and ultimately a stewardship plan that would guide the tribal natural resources department in their lake sturgeon restoration. This group had representatives from many different sections of the tribal community; men and women, elders and youth, artisans and pipe carriers.

Rather than just having biologists determine restoration strategies, this group would provide a tribal voice in the management direction. This "voice" was an amalgamation of cultural, biological, political, and social elements, all being important and often indistinguishable from each other. The biologists of the group recognized early on that an exciting part of developing a sturgeon plan was using sound biological principles to meet objectives that were not necessarily or exclusively biological. For instance, the first goal of the stewardship plan was "Restore the harmony and connectivity between nmé and the Anishinaabek and bring them both back to the river." Bringing the sturgeon back to the river was an obvious biological element; however, restoring harmony and connectivity between sturgeon and people was steeped in the cultural and social realm.

Within these Cultural Context Group meetings, we observed that the depth of the relationship between the sturgeon and the tribe would bring a unique management perspective. Each "meeting" began with a ceremony, and the conversation was held over a feast, including wild rice soup and fry bread. After almost two years the stories, the friendly dialogue, and the vision shared at these meetings would produce a plan that would guide the LRBOI's sturgeon work for the next seven generations. The four goals of the plan were these:

- Restore the harmony and connectivity between nmé and the Anishinaabek and bring them both back to the river
- Restore the nmé and reclaim the environment on which it depends for future generations of nmé and Anishinaabek in perpetuity
- Emphasize strategies that promote natural reproduction and a healthy watershed
- Protect tribal sovereignty and treaty rights

Medicine and Drift Nets

One of the most challenging species to rehabilitate in the Great Lakes may be nmé, largely due to a life history that is extreme when compared to other freshwater species. Probably the most daunting realization for me as a nmé biologist is that the management actions that we implement today will not be fully realized within my career because nmé mature so slowly and may spawn for the first time only after 10–20 years of life (Auer 1996). The first challenge in implementing a restoration plan was to fully understand the status of the sturgeon population. One of the best ways to determine the status is to capture and count the newly hatched larvae. A larval drift survey may be one of the more challenging and demanding surveys that biologists conduct within the Great Lakes Basin, and "larval drift" therefore is a phrase that is banned from the LRBOI natural resources department once the surveys are completed in the spring because of the exhaustion and stress that these surveys cause (this is only a partial joke!).

Why the disdain for drift surveys? Unfortunately for fisheries staff, nmé larvae (fry) can most effectively be captured at night. After nmé hatch they remain buried in the gravel for a few days; once they have absorbed most of their yolk sac they drift downriver starting just after sunset through the early morning hours. To capture nmé larvae fisheries staff need to become nocturnal, face the rain, snow, and ice, hike through the woods, and wade into chest-deep, swift-moving rivers with headlamps to set anchors and check drift nets. This is done every night for many consecutive days.

When we started conducting these surveys in 2002, no lake sturgeon larvae had been captured in any Lake Michigan tributary including the Big Manistee River even after two years of drift surveys had been conducted by a university. This was alarming because it indicated that either no nmé were successfully reproducing or only a few were surviving. Either way the prospects appeared bad. We decided to move our drift study site closer to the expected spawning area than the past researchers to increase the probability of capturing nmé. Three LRBOI staff, Mark Bowen, Stephanie Ogren, and I, began heading down to a desolate location on the bank of the Big Manistee River at night to attempt capture of young nmé. We set out four drift nets into the fast-flowing water, collected our catch every hour, and meticulously went through the samples. For the first few nights we caught almost everything in the river but sturgeon—thousands of sucker larvae, aquatic insects, lamprey amocetes, salmon, and trout fry.

We were frustrated and disheartened until one evening Mark Bowen told us that we needed to follow a teaching that had been passed on to him. Both Stephanie and I watched as he pulled out his medicine pouch, put tobacco into the water, and prayed that the grandfather fish would allow itself to be captured. I mention this because it is an example of trying to achieve the first goal within the plan, bringing the sturgeon and the people back to the river (restoring the fish/human relationship). This also exemplifies the integration of culture and biology—the tobacco floating upon the river right next to drift nets.

That night I remember anticipating our first net pull to see if we would find a nmé. As I poured my sample slowly into a white tray I scanned for signs of a swimming nmé with my flashlight. As I looked at dozens of frantically swimming sucker larvae I saw something different, a grayish fish with a blunt head that was swimming entirely differently. For a moment I didn't know what species of fish it was. I had seen hundreds of sturgeon larvae before on the Sturgeon River (Barage County, Michigan) with my graduate advisor, Nancy Auer, but this time I was the one responsible for identifying the sturgeon, and this fish was smaller than any one I had observed before.

I didn't realize it at the time but I was holding my breath, and my two partners, both aware of this, had stopped going through their samples and were intently watching me. When I looked up at them with a grin on my face, they both knew we had a fish, a sign that there were still reproducing sturgeon in the Big Manistee River. Mark leaned over our table, over the little sturgeon, and gave me a hug. After the fish was measured Mark carefully took the fish in his hand and released it back to the river. The connection between people and sturgeon had begun. This little sturgeon, and the few others we captured that year, demonstrated there was a small amount of natural reproduction taking place and also gave us an idea that would eventually be applied across the Lake Michigan Basin for restoring sturgeon.

Bringing Back the Sturgeon

Tribes are not often recognized for the large amount of fishery research and restoration they do. Each year tribes in the Great Lakes conduct thousands of hours of biological assessments and numerous habitat restoration projects and bring millions of dollars into fishery improvements, such as fish stocking, improved road-stream crossings, bank stabilizations, and dam removals (USDOI 2007; Snyder et al. in preparation). This work benefits not only the tribal members but each and every person who enjoys the natural resources of the area. The LRBOI began dedicating hundreds of hours to sturgeon research after capturing the first sturgeon larvae. We captured eggs that in turn identified spawning locations, we found juvenile fish that demonstrated survival from larvae, and we began to determine what habitat conditions were present and needed for spawning lake sturgeon. With this information, we could begin considering management approaches for restoring the nmé.

Since 2002 a dedicated group of scientists has gathered every two years to discuss lake sturgeon rehabilitation strategies. One of the critical topics discussed at the first meeting was the unique genetic arrangements and structure that was observed in remnant populations of nmé across the Great Lakes. By taking a tissue sample from a nmé (a small piece of a fin) researchers were able to assign quite accurately the river in which that particular fish had originated (Welsh et al. 2010; Welsh et al. 2008; DeHann et al. 2006). Genetic structuring indicated that sturgeon were philopatric, meaning that the spawning fish returned faithfully to the same river that they themselves originated (DeHaan et al. 2006). This imprinting of fish to a particular area for spawning is exhibited in other species as well, notably salmon, and changes the way we should look at management of these fish. This meant something very important to those at the meeting and for those making management decisions for nmé; this genetic structuring (and imprinting) needed to be maintained because of important evolutionary traits that could be unique to each of the populations.

The LRBOI started to develop a strategy for increasing sturgeon abundance in the Big Manistee River. One strategy historically used in sturgeon restoration was rearing fish in an off-site hatchery often far away from the river where the fish would be stocked. However, during the LRBOI Cultural Context Meetings we had discussed this strategy, and many of the participants were not supportive of the idea based on cultural values. It was clearly communicated that the sturgeon was a grandfather fish and part of the Big Manistee River watershed at all stages of life for a reason. The risk of altering their behavior by being reared in water from an off-site hatchery was not acceptable. The stewardship plan was clearly guiding us to keep the fish in the river at all times: "The Creator put the nmé in the Big Manistee River. The nmé and the rivers they use are part of our sense of place. The Creator put us here where the nmé return. We are obliged to remain and protect this place."(LRBOI 2008).

The group believed that maximizing the chances of imprinting was essential, not only for the Big Manistee River sturgeon but also for the small surrounding populations where straying of Big Manistee River fish could adversely impact those populations as well. As Don Stone, a tribal elder, put it, "These fish were here when our ancestors were here," and we needed to make sure that they would come back as they always had. The biologists needed to determine a strategy that would accommodate the cultural and biological perspectives related to sturgeon restoration. The second question I posed above, "Would a tribal *management* approach be different from that of the other agencies?" has a simple answer—yes. There are often different management goals based on the unique culture of the tribes that in turn create different strategies toward management.

I find it interesting that at the same time sturgeon managers from across the Great Lakes were questioning traditional fishery management techniques for restoring sturgeon, tribal people were coming up with similar conclusions based on culture and biology. It demonstrated to me the great conservation potential that may be gained by including multiple perspectives and developing a shared knowledge where the outcome (which will be described below) will often be much richer and innovative than if only one perspective was included (Natcher, Davis, and Hickey 2005). By combining these perspectives there is much more than sturgeon restoration being accomplished. Jay Sam and Art deBres describe this philosophy as, "We are not introducing, we are rehabilitating . . . we are assisting and saving our mother, grandfather and cousin." (LRBOI 2008).

A New Approach to Sturgeon Management: Streamside Rearing

In 2003 the LRBOI decided to design and operate a streamside rearing facility to rear lake sturgeon (Holtgren et al. 2007). The larvae that were captured during drift surveys would be protected for a few months inside the rearing facility and released back into the river (plate 12). Even though streamside rearing had been used successfully for salmon, trout, and walleye (Dupuis and Dominy 1994; Steward 1996), it had never been applied to sturgeon. The Big Manistee River system provided its own unique challenges for keeping the streamside-rearing facility running effectively. The challenge for us was to create a system that pulled the water from the river without having the rearing tanks inundated with the silt and sand that the river carried, especially during high flow events. The system also had be cost-effective, incorporate genetic conservation, and address the concerns of imprinting and spawning site fidelity (Holtgren et al. 2007).

The streamside rearing facility was built inside a cargo trailer for mobility (plate 13). It is pulled by truck to a forested area along the banks of the Big Manistee River each spring and put into storage each fall. From the outside it appears to be indistinguishable from one you would see on the road except for the large tribal logo and black lettering that says, *Nmé Kooginaawsawin Koh-ge-now-sa-win*, which means "The sturgeon home, where children are raised" in Anishinaabemowin (the language of the Ottawa). However, the inside is complete with water quality monitors, alarms, and safety systems. The water is pumped into the facility through 100 meters of underground piping and enters a set of mechanical sediment filters, which remove a portion of the silt and sand load. The water is then forced into a large reservoir above the trailer and gravity-fed through the sidewall into the tanks and immediately drains back into the river. Because of the remote location of the rearing facility, a safety system was necessary in case of a malfunction. We have a notification system that activates a phone when the water decreases to a low flow rate or stops and practically calls everyone in the department until someone checks to see what the problem may be. There have been nights when someone receives a call at two or three in the morning and our department phone tree lights up. No one rests until the fish are safe.

Our larval collection procedure and the streamside rearing approach were appealing for many reasons. First, by capturing larvae for rearing we were allowing the spawning sturgeon to continue their natural process. Instead of capturing the prespawn fish and "stripping" their eggs and sperm, the sturgeon were selecting how they would mate and therefore continue the unique genetic structuring found within the population. Also, we decided to only collect 10 percent or less of the drifting larvae to ensure that if something did happen to the fish we were rearing, an adequate number of wild fish could provide that year's production.

In a strict sense of the word, we were not necessarily even "stocking" fish but augmenting the population by removing fish already within the population and simply increasing their prospects of survival. When sturgeon are larvae they are quite vulnerable to many sources of predation. Many of these predators are relatively new to the watershed (intentional and unintentional releases) and are species the sturgeon has not necessarily developed defense strategies against. By rearing collected larvae we could also increase the chances of imprinting by keeping the fish in their own river water throughout their early life.

Bringing Both Back to the River

The week leading up to the release of streamside-reared fish is hectic, and we are busy being scientists. Each fish is marked with a small internal tag so if it is later captured it can be identified and we can evaluate if the program is meeting its goals and objectives. Tissue samples are collected, including a small snip of the tail fin, to better understand the genetic makeup of the population and to ensure that what we release is representative of that in the wild. We attach small radio transmitters to a portion of the fish to monitor how the reared fish behave compared to their wild counterparts. After those activities, the week is no longer about science but about people and fish coming together next to the banks of the river. Marcella Leusby described this concept by offering, "When putting the sturgeon back in the river, I felt it was one of the most meaningful acts the LRBOI had done. It was very emotional." (LRBOI 2008).

The nmé release is a celebration, a celebration with drumming and dancing,

Figure 1. Drummers playing at the nmé release ceremony.

Figure 2. Young people releasing young sturgeon for future generations.

Figure 3. Jimmie Mitchell singing a song while nmé are released.

Figure 4. Mark Bowen with a transmittered nmé about to be released.

recognizing invaluable partners and finally commencing with a pipe ceremony. After the pipe, the nmé are taken from the raceways, carried overland in five-gallon buckets, and then released one by one, by hand, into the river. After the fish are released, a crowd of a couple hundred people stand silently for a few moments and begin to talk about the nmé and about how they will be ready to celebrate again next year. The gathering of people is a mixture of the watershed community encompassing both tribal and nontribal people, where the differences among cultures are celebrated and the similarities are apparent.

Bringing nmé and people back to the river is now a goal of the watershed community. Yes, the tribe largely funds the program (with support from the U.S. Fish and Wildlife Service) and provides the expertise, but the nontribal community provides tremendous support. The day when we release nmé into the river is just as much about human ecology, how humans fit into the natural and social environments. It is at this type of celebration that I understand we are once again learning humans are part of the ecosystem, not an organism somehow separated. By putting our hands in the water and having the young nmé slowly swim away, we are seeing ourselves in a different way.

After rearing sturgeon for over seven years, the stewardship plan of the LRBOI is beginning to be realized. Each year adds new memories and stories to the rich relationship between humans and nmé. The beginning of nmé rearing begins with a simple act, someone taking their hand and gently guiding the newly captured nmé larvae into a holding tank to be transported to the streamside rearing facility. The end of nmé rearing also begins with a hand gently guiding the larger nmé (around 250 mm TL [10 inches]) back into the Big Manistee River.

Final Thoughts

I stated at the beginning of this chapter that a goal of the stewardship plan was to promote tribal sovereignty and treaty rights. Tribal sovereignty allows tribes to conduct projects like nmé restoration in the Big Manistee River, to bring people back to the river, to appreciate where we come from, and ultimately to understand each other. In this example, sovereignty is not some scary prospect where there is misunderstanding, confusion, and even anger between tribal and nontribal people. This is about harmony.

REFERENCES

Auer, N. A. 1996. Importance of habitat and migration to sturgeons with emphasis on lake sturgeon. Canadian Journal of Fisheries and Aquatic Sciences 53(S1):152–160.

———. 1999. Lake sturgeon: A unique and imperiled species in the Great Lakes. *In*: Great Lakes Fisheries Policy and Management: A Binational Perspective. W. W. Taylor and C. P. Ferreri, eds. Michigan State University Press.

Berkes, F. 2009. Evolution of co-management: Role of knowledge generation, bridging organizations and social learning. Journal of Environmental Management 90:1692–1702.

Busiahn, T. R. 1989. The development of state/tribal co-management of Wisconsin fisheries. *In*: Cooperative Management of local fisheries: New directions for improved management and community development. E. Pinkerton, ed. University of British Columbia Press.

Coffey, W. and R. Tsosie. 2001. Rethinking the Tribal Sovereignty Doctrine: Cultural sovereignty and the collective future of Indian nations. Stanford Law & Policy Review 12:191–202.

DeHaan, P. W., S. T. Libants, R. F. Elliott, and K. T. Scribner. 2006. Genetic population structure of remnant lake sturgeon populations in the upper Great Lakes Basin. Transaction of the American Fisheries Society 135:1478–1492.

Dupuis, T., and L. Dominy. 1994. Introduction to satellite rearing: Installation and operation manual. Atlantic Salmon Federation.

Great Lakes Indian Fish and Wildlife Commission (GLIFWC). 2007. A guide to understanding Ojibwe treaty rights.

Harkness, W. J. K., and J. R. Dymond. 1961. The lake sturgeon. Ontario Department of Lands and Forests, Fish and Wildlife Branch.

Henquinet, J. W. and T. Dobson. 2006. The public trust doctrine and sustainable ecosystems: A Great Lakes fisheries case study. New York University Environmental Law Journal 14(2): 322–373.

Holtgren, J. M., S. A. Ogren, A. J. Paquet, and S. Fajfer. 2007. Design of a portable streamside rearing facility for lake sturgeon. North American Journal of Aquaculture 69:317–323.

Kimmerer, R. N. 2000. Native knowledge for native ecosystems. Journal of Forestry 98:4–9.

LRBOI (Little River Band of Ottawa Indians). 2008. Nmé (Lake Sturgeon) stewardship plan for the Big Manistee River and 1836 reservation. Natural Resources Department, Special Report 1.

Mattes, W. P., and N. Kmiecik. 2006. A discussion of cooperative management arrangements within the Ojibwa ceded territories. *In*: Partnerships for a common purpose: Cooperative fisheries research and management. A. N. Read and T. W. Hartley, eds. American Fisheries Society Symposium 52.

McClurken, L. M. 2009. Our people, our journey: The Little River Band of Ottawa Indians. Michigan State University Press.

Natcher, D. C., S. Davis, and C. G. Hickey. 2005. Co-management: managing relationships, not resources. Human Organization 64(3): 240–250.

Nielsen, L. A. 1999. History of inland fisheries management in North America. *In*: Inland fisheries management in North America. 2nd ed. C. C. Kohler and W. A. Hubert, eds. American Fisheries Society.

Notske, C. 1995. A new perspective in aboriginal natural resource management: Co-management. Geoforum 26(2):187–209.

Rettig, B. R., F. Berkes, and E. Pinkerton. 1989. The future of fisheries co-management: A multidisiplinary assessment. *In*: Cooperative management of local fisheries: New directions for improved management and community development. E. Pinkerton, ed. University of British Columbia Press.

Runstrom, A., R. M. Bruch, D. Reiter, and D. Cox. 2002. Lake sturgeon (*Acipenser fulvescens*) on the Menominee Indian Reservation: An effort toward co-management and population restoration. Journal of Applied Ichthyology 18:481–485.

Salmon, E. 2000. Kincentric ecology: Indigenous perceptions of the human-nature relationship. Ecological Applications 10:1327–1332.

Schoolcraft, H. R. 1970. Narrative journals of travels through the northwestern regions of the U.S. extending from Detroit through the great chain of American lakes to the sources of the Mississippi in the year 1820. M. L. Williams, ed. Arno Press and the New York Times.

Steward, C. R. 1996. Monitoring and evaluation plan for the Nez Perce Tribal Hatchery. U.S. Department of Energy, Bonneville Power Administration, Report 36809-2.

Tody, W. H. 1974. Whitefish, sturgeon, and the early Michigan commercial fishery. *In*: Michigan Department of Natural Resources. Michigan fisheries centennial report, 1873–1973. Michigan Department of Natural Resources, Management Report 6.

USDOI (U.S. Department of the Interior). 2007. Fishery status update in the Wisconsin treaty ceded waters. 4th ed. Bureau of Indian Affairs.

Welsh, A., R. Elliott, K. Scribner, H. Quinlan, E. Baker, B. Eggold, J. M. Holtgren, C. Krueger, and B. May. 2010. Genetic guidelines for the stocking of lake sturgeon (*Acipenser fulvescens*) in the Great Lakes Basin. Great Lakes Fishery Commission Special Publication. 2010-01.

Welsh, A., T. Hill, H. Quinlan, C. Robinson, and B. May. 2008. Genetic assessment of lake sturgeon population structure in the Laurentian Great Lakes. North American Journal of Fisheries Management 28:572–591.

BRENDA ARCHAMBO

Sturgeon for Tomorrow

ONE OF THE MOST COMMON QUESTIONS I'VE BEEN ASKED OVER THE YEARS IS, "How did you become so involved with the lake sturgeon?"

I grew up logging hundreds of hours in an ice shanty on Black, Burt, and Mullett lakes in northern Michigan's Cheboygan County. My late father and brother Dock McCall and James McCall taught me to fish from an early age. I am so blessed they took me into the wilderness and upon our great waters, and for letting me roam free and discover earth's natural treasures, beauty, and peace. I have never lost the wanderlust for life they and my late mother have instilled in me.

In the mid-1990s, I traveled to southern Michigan to spend time with my grandfather Roy Naugle. Grandpa farmed his whole life. He worked hard from daylight till dark eleven months out of the year. His passion was to spend the other month ice fishing on Black, Burt, and Mullett lakes.

During my visit, I knew Grandpa was dying. We cried, we laughed, and we talked about sturgeon. Grandpa held my hand, looked me in the eye, and said, "You do what you need to do to keep the sport alive."

I was with my grandfather the first time I saw a sturgeon. I was about six years old, and I will never forget that moment. I knew then that there was something extraordinary about this creature.

Grandpa and I were sitting in a fish shanty on Burt Lake one February morning in the late 1960s. Suddenly a commotion came from a shanty nearby. We threw the door open to see what was going on. There lay a huge sturgeon on the ice, bigger than I had ever seen. A few dozen anglers from surrounding shanties began to gather around the sturgeon. We trudged through the snow to see the fish close up. The spirit in the air was powerful. I remember looking into the eye of the sturgeon. It amazed me. The diamond shape of the pupil reminded me of pictures I had seen of dinosaurs. That moment and the excitement of that day on the ice with Grandpa have stayed with me ever since. And so have the traditions associated with the sturgeon in our region.

Onaway, a quaint community five miles south of Black Lake, is called the Sturgeon Capitol of Michigan. There the Black Lake Sturgeon Shivaree debuted in 1963. The Shivaree is a family fun weekend ice-carnival on Black Lake celebrating sturgeon.

A large tent is set up on the ice. It is heated and a central meeting place for everyone to gather, socialize, and register for events. Numerous events for all ages include snowmobile races, ice skating, cross-country skiing, adult and children's games, food and outfitter vendors, fishing contests, and of course, plenty of beer and a sense of place and community.

A sturgeon king or sturgeon queen is crowned for harvesting the largest sturgeon. Area businesses thrive during Shivaree weekend and during ice fishing. Getting out and enjoying the great outdoors is a perfect opportunity for families to gather and break the monotony of cabin fever.

Black Lake has long been recognized by local residents and conservation groups for its natural resources and, is, we believe, a key aquatic biodiversity site of the Great Lakes ecoregion.

In addition to large kettle lakes, large forested areas, and an expansive network of streams and wetlands, the Black Lake watershed is home to a variety of threatened and endangered aquatic species, including not just the iconic lake sturgeon but also the Michigan monkeyflower and the Hungerford's crawling water beetle. Several wetlands also provide important nesting habitat for rare birds such as the bald eagle, the common loon, and the great blue heron.

Every spring for decades, locals have paddled the Upper Black River to see lake sturgeon spawning in all their glory. Families camp and picnic at several of the known spawning sites, where you can see dozens of the majestic sturgeon spawning in a short span of the river.

By the 1990s, it became clear there were poachers on the river, especially at night. Poachers illegally harvested the sturgeon in the spring spawning run, selling the roe and smoked fillets on the black market. Poaching was reducing spawning stocks and hindering natural reproduction.

This practice has a long history. Beginning in the middle to late 1800s, North Americans became aware of the value of sturgeon. Europeans considered caviar a delicacy and so the demand for sturgeon exploded. In addition to caviar, sturgeon were harvested for a number of purposes: sturgeon meat was delicious, especially smoked; the skin was tanned for leather; and the swim bladders were used for isinglass, a high-quality gelatin used for pottery cement, waterproofing, and clarifying wine and beer.

Poaching is and was a family tradition. Some family members who poached them in the past are still bent on taking them, primarily for caviar and their meat. Poachers created an underground network whereby the roe would be bootlegged to Chicago and then to New York, where it was sold on the black market. Today, the Convention on International Trade of Endangered Species (CITES) monitors the import and export of sturgeon, while the U.S. Fish and Wildlife Service is the watchdog for international trade of threatened and endangered animals including sturgeon.

Legal harvest of sturgeon is a family tradition, too. In our area there are two, three, and in some instances four generations of avid sturgeon anglers. Ice fishing on area lakes is deeply entrenched in the local culture. Sturgeon is a trophy fish, and fishing for them is indeed a hunt of a lifetime. Fish decoy carvers and spear craftsmen handcraft one-of-a-kind spears and award-winning decoys. Many ice anglers became deeply concerned about the future of their sport of spear fishing through the ice. I was committed and passionate about working to protect this tradition and to ensure a self-sustaining sturgeon population.

In 1995 I learned the Michigan Department of Natural Resources (DNR) was reviewing statewide sturgeon regulations. In 1997, the DNR conducted a lake survey to assess the lake sturgeon population in Black Lake. The estimated adult spawning population was 550, and the entire population was thought to be about 1,300. DNR recommended closing the sturgeon harvest completely in all three lakes. We believed poaching was largely responsible for the sturgeon's decline and that the more recent survey methodology was not standardized with the survey of 1975.

It was believed poachers would take many more sturgeon illegally than were taken in legal harvest. It was commonplace. Everyone did it. Fathers of today's poachers had done it. Their grandfathers, uncles, and cousins had done it. But at several public meetings in 1996–1997 conducted to receive community input on the proposed regulation changes, the state's position was that the sturgeon population was not reproducing because of fishing pressure.

We conducted a statewide outreach campaign, and with the help of local anglers and businesses, raised enough money to purchase several larval drift nets to loan to the DNR to assess if there was or was not in fact natural recruitment in the river. Studies have determined there is natural reproduction occurring in the river.

The outreach campaign informed ice anglers of the proposed sturgeon regulation changes. Included in the outreach materials were goals and recommendations from the State of Michigan Lake Sturgeon Rehabilitation Strategy (LSRS). For water bodies with sturgeon populations of 500 or more breeding adults (Black Lake's estimates were 550 breeding adults), there could be a 3 percent harvest for an expanding population, or 6 percent harvest for a sustaining population. Black Lake fell within these parameters. Mullett and Burt Lakes had limited data on sturgeon population assessments.

Thirty days into the outreach campaign, we received over 1,300 petition signatures from anglers supporting the goals and recommendations of the LSRS. The overarching goal of the campaign was to determine if there was enough interest in lake sturgeon rehabilitation, and the tenacity to save the sturgeon and save the sport that had become deeply entrenched in the local culture. Indeed there was, and is today!

Regardless of the regulation changes, poaching needed to be reduced. Here we had a valuable, rare, living fossil, and not enough was being done to protect it. Local DNR officials, who knew for years there was a huge poaching problem, had set up sting operations that were largely ineffective.

So members of the community and I began tossing out ideas. In 1999, the Black Lake Chapter of Sturgeon for Tomorrow (SFT) incorporated as a 501(c)3 nonprofit organization. The mission: To assist fisheries managers in the rehabilitation of lake sturgeon. That year, we coordinated the first annual Sturgeon Guarding Program, modeled after the successful program at Wisconsin's Lake Winnebago.

We began recruiting volunteers to stand watch and camp along the Black River during the spawning run to protect the sturgeon from would-be poachers. Boy Scout troops, Vietnam veterans, volunteer off-duty National Guardsmen, sportsmen and women, retirees, and people from throughout the Great Lakes Basin and Canada volunteered to patrol the area around the clock all year round. Today, about 350 volunteers contribute 3,800 hours along the river each year as a deterrent to poachers.

The sturgeon migration is an amazing spectacle. We can identify sturgeon that have been tagged before when they come up the river. Although the literature says this return should happen every two to four years, we see some males returning to the river every year, and they are contributing to reproduction. We have also seen some previously tagged females. Some of these are just under seven feet in length.

The annual spawning run is a wonder of nature at its finest. It is spectacular to see the sturgeon in all its glory. You look down from the cliff, look up and see eagles, and hear the rippling of the water. When sturgeon are spawning, you can hear them thrash and literally feel the ground shake.

Since so many people were out on the river during spawning, it made sense to conduct research, compile data, and learn more about the reproductive capacity and early life history of the spawning population. Little was known about the spawning population of the Black Lake sturgeon. In 2000, Central Michigan University took on the project. In recent years, Michigan State University and the state Department of Natural Resources have directed the research.

Researchers net and tag the adults to collect population data, and small fin samples are analyzed to determine genetic diversity. Larval sampling has quantified natural reproduction. The naturally produced larvae are collected, transferred, and reared at the streamside-rearing facility, then released after three months. Eggs and milt are also cultured to maximize production. There have been over 41,000 sturgeon fingerlings released into Black, Burt, and Mullett lakes since 2000. The Tower-Kleber Limited Partnership owns the dam and property on which the hatchery operates. The DNR leases the facility. With the leadership of Huron Pines and the hands of dozens of SFT volunteers, habitat improvements and streamside interpretive signage have been implemented to preserve some of the last known spawning habitat in northern Michigan.

Guided eco-tours to see the spawning sturgeon and hatchery tours have proven to be highly popular. Meanwhile, SFT and collaborators are developing interpretive programming, including a visitors pavilion near the hatchery to expand outreach and education. Our annual banquet, golf scramble, memberships, and contributions primarily fund these initiatives, and grants have funded university research.

Today, Sturgeon for Tomorrow has a vision of delisting the sturgeon from threatened status in the state and creating a world-class fishery. Sturgeon populations must be closely assessed and managed. States, tribes, universities, and nongovernmental organizations should collaborate to reestablish healthy, sustainable populations. Management plans will be developed with input from all stakeholders and then implemented. Since 2010, the LSRS has been under revision. Rehabilitation efforts, including biological research and strict regulations with stiff penalties for violators, should ensure there will be sturgeon populations for future generations to enjoy. But while sturgeon are being aided by these actions, people must remember that since sturgeon are such slow-growing, long-lived fish, it will be many years before we see populations restored to the levels resembling anything experienced in the past. Through our outreach programs, we will continue to educate, engage, and mobilize diverse constituencies to treasure the majestic lake sturgeon and our natural world and not take them for granted. We must all work together to keep our great lakes—great.

In my view, conservation embraces values that spring from an early and profound childhood experience in nature. Sense of place deepens. The common thread of my childhood is that I spent a lot of time outdoors because that's where the things that interest me live. It's where I developed many of my core values. Understanding the threats to the places I love and where I'd gained self-assurance made me a conservationist and convinced me that connecting people with nature, especially children, is one of the major tasks to winning this great war, and standing up to crimes against nature. We must learn how best to live sustainably and in harmony with nature, and to pass this rich heritage onto future generations.

The sturgeon is the oldest and largest fish in the Great Lakes, the elder statesman of Michigan fish species. It is an ancient fish. Sturgeon are our ancestors, a living fossil. Their sheer existence and wonderful display every spring in the river, and in the winter the ice-fishing heritage, have become deeply entrenched in our culture. Like the sturgeon, we are inextricably an element of the ecosystem.

HOLLY MUIR AND TRENT M. SUTTON

The Relationship between Lake Sturgeon Life History and Potential Sensitivity to Sea Lamprey Predation

THE LAKE STURGEON *ACIPENSER FULVESCENS*, A SPECIES NATIVE TO THE LAUREN-
tian Great Lakes, has a unique life history. Like other sturgeons, lake sturgeon are a
slow-growing, long-lived species with delayed maturation; first spawning for males
typically occurs between ages 12 and 15, while females become mature between ages
18 and 27. In addition, lake sturgeon spawn intermittently, with females spawning
only once every four to nine years and males spawning every one to three years
(Roussow 1957; Scott and Crossman 1973; Fortin, Dumont, and Guénette 1996;
Bruch 1999; Bruch, Dick, and Choudhury 2001). Although these life-history traits are
advantageous for buffering against extreme environmental conditions, they increase
susceptibility to human-induced mortality and the negative effects of aquatic invasive
species (Hay-Chmielewski and Whelan 1997; Auer 2004).

Despite supplemental stocking, efforts to improve water quality, and permitted
harvest reductions, lake sturgeon populations have been slow to recover from their
imperiled state throughout the Great Lakes basin (Welsh et al. 2008). This slow
recovery is not surprising, considering the great age-to-maturity characteristic of
lake sturgeon. However, another potential factor that may be currently impeding lake
sturgeon rehabilitation is aquatic invasive species. For example, the invasive parasitic

sea lamprey *Petromyzon marinus* preys on and kills lake sturgeon in confined tank experiments (Patrick, Sutton, and Swink 2009). The objective of this chapter is to review the effects of aquatic invasive species on Great Lakes lake sturgeon populations, with an emphasis on the potential effects of sea lamprey predation on both the decline and slow rate of recovery of lake sturgeon.

Aquatic Invasive Species in the Great Lakes

The introduction and establishment of undesirable, nonnative plant and animal species is one of the greatest threats to the future of the Great Lakes and to species and community conservation worldwide (Finster 2007; Thresher 2008). The Great Lakes have been subject to invasion by nonnative species since European settlement (Mills et al. 1993). Development of the basin, including timber harvest, agriculture, hydropower development, and canal construction, as well as invasion of marine organisms, fish stocking, and overharvest, have resulted in rapid changes in what was once a simple, slowly evolving ecosystem (Fetterolf 1980).

At least 182 nonindigenous species have been introduced into the Great Lakes since 1840, with over 40 percent of these invaders occurring after the opening of the St. Lawrence Seaway in 1959 (Ricciardi 2006). The Great Lakes have the highest rate of nonnative species invasions and introductions recorded in any freshwater ecosystem. Since 1960, the invasion rate is estimated to be 1.8 species per year, equivalent to one new invader being discovered every 28 weeks. Though not all nonnative species jeopardize the Great Lakes ecosystem (some, such as the Pacific salmon, have been intentionally introduced to support commercial or recreational fisheries), those that are considered to be injurious are capable of inflicting significant damage to the environment and the economy (Finster 2007). The degree to which native fish and their habitats are affected by injurious aquatic invasive species depends on the life-history traits of invading and of native species, as well as the ability of the ecosystem to withstand change (Ricciardi 2006). To date, little is known about the injurious effects of aquatic invasive species on lake sturgeon.

The Voracious Sea Lamprey

Of all the aquatic invasive species thought to impact lake sturgeon abundance, the sea lamprey poses the most serious threat. The sea lamprey is similar in appearance to the American eel *Anguilla rostrata*, but lacking a jaw and paired fins, and having seven pairs of gill pouches instead of the usual gill structure of bony fishes. The life cycle of the sea lamprey involves distinctively different larval and adult feeding phases. Whereas larval lampreys, called ammocoetes, feed primarily on organic detritus, adults in their predacious stage feed on the blood of other fishes (their hosts) (Lowe, Beamish, and Potter 1973; Sutton and Bowen 1994). After spending an extended larval phase (3 to 10 or more years) buried in the sediment of streams, sea lamprey larvae metamorphose into parasites, developing eyes and teeth, and enter the lakes to feed. The period of metamorphosis between these two feeding phases is called "transformation," and lampreys at this stage are referred to as "transformers," or macropthalmia because of their large eyes (Applegate 1950).

Each summer and fall, one age group of parasitic sea lamprey is actively feeding. The next spring, this group migrates back to tributary streams to spawn and die. When feeding on a fish, sea lamprey attach to their host with their suction-cup-like mouth, called an oral disc, which is armed with concentric rows of sharp teeth. In the center of their mouth is a sharply toothed tongue, which is used to rasp holes in the flesh of their prey to feed on its blood and tissue. Sea lampreys are also equipped with a buccal-gland system that secretes anticoagulant, called lamphredin, to thin the blood and tissues of its host. Because the average sea lamprey grows from an approximately 10 g transformer to a 150 g adult in only 18 months, one can imagine how much blood must be consumed and, in turn, how many fish a lamprey must kill, during its parasitic life stage. A single sea lamprey can kill an estimated 18 kg of fish or more during its life (Farmer 1980).

Origins of the Sea Lamprey in the Great Lakes

The sea lamprey is native to the Atlantic Ocean and; although its origin in the Great Lakes, Lake Champlain, and the large lakes in New York has been long debated, landlocked freshwater populations also exist in these lakes (e.g., Bryan et al. 2005). Most proponents support one of three zoogeographic scenarios explaining the possible origins of freshwater sea lamprey populations. Freshwater populations in Lake Ontario and Lake Champlain may have arisen from (1) natural migrations from marine sources in the St. Lawrence River or the Atlantic coastal drainages; (2) as a result of historical bio-geographic events such as periods of marine submergence (i.e., the incursion of the Champlain Sea [11,500–13,000 years ago]); or (3) by recent invasion from the Hudson River through canals built for transportation purposes in the nineteenth and twentieth centuries (i.e., Erie and Champlain canals) (Bryan et al. 2005).

Proponents of the theory that sea lamprey entered Lake Ontario by way of natural migration maintain that this species is indigenous to Lake Ontario and its tributaries (Hubbs and Lagler 1947; Bigelow and Schroeder 1948; Rostlund 1952; Lawrie 1970; Scott and Crossman 1973; Bailey and Smith 1981; Smith 1985; Daniels 2001; Waldman et al. 2004). These supporters hold that the discontinuous distribution between the freshwater populations in the New York Finger Lakes and the Hudson River population is evidence that sea lamprey are native to Lake Ontario and its tributaries (Mills et al. 1993). However, DeKay (1842) found sea lamprey as far upstream in the Hudson River as Albany, New York (Mills et al. 1993). Modeling results of Bryan et al. (2005) also support the hypothesis that sea lamprey migrated to the freshwater lakes via the St. Lawrence River. For example, the presence of a rare genetic composition (allele) in landlocked Lake Ontario and Lake Champlain populations is evidence that populations in these lakes have been separated from the putative North American progenitor populations on the Atlantic Coast for a considerable period of time. Another argument supporting the hypothesis that lampreys are native to Lake Ontario is that the presumed poor water quality conditions of nineteenth-century navigation canals, in combination with the impediment of numerous locks, make canals an unlikely vector for lamprey dispersal (Daniels 2001).

The conclusions regarding endemnicity of the Lake Ontario population may be overstated (Eshenroder 2009). Because observations of sea lamprey in Lake Ontario were not reported until the 1830s, a second belief is that the sea lamprey is not native to Lake Ontario, but rather invaded the lake in historical times after the construction and opening of the Erie Canal in the early 1800s (Aron and Smith 1971; Emery 1985; Mandrak and Crossman 1992; Mills et al. 1993; Smith 1995). Most support for the invasion-by-canal hypothesis lies in the disbelief that sea lampreys could have been historically present, but gone unobserved in Lake Ontario until after the construction of the Erie Canal (Eshenroder 2009). Even if the sea lamprey is indeed native to Lake Ontario, Niagara Falls served as an insurmountable natural barrier that effectively prohibited the migration of the parasite into the remaining Great Lakes.

It is well established that the construction of the Welland Ship Canal in 1829 provided the necessary thoroughfare for the sea lamprey to gain access to Lake Erie, and eventually, the upper Great Lakes. The sea lamprey now inhabits all five of the Great Lakes, where it is considered by U.S. and Canadian management agencies to be an invasive species. Even if sea lampreys are native to Lake Ontario, they have proven to be capable of exhibiting characteristics of invasive species when predator-prey dynamics are disturbed.

The Effects of Sea Lamprey Predation on Fish Communities

Sea lamprey were first reported in Lake Erie in 1921, Lake St. Clair in 1934, Lake Michigan in 1936, and Lake Huron in 1937 (Dymond 1922; Trautman 1949; Smith and Tibbles 1980). The rapids and Soo Locks at the lower end of Lake Superior apparently hindered their invasion; the first confirmed record in Lake Superior was a parasite near Isle Royale in 1946 (Applegate 1950). Sea lampreys and lake trout *Salvelinus namaycush* coexisted in Lake Ontario for a long time. Annual production of lake trout in Lake Ontario did not appear to decline abruptly as it did in the upper Great Lakes; annual production from 1867 to 1940 averaged 164.7 MT and declined to only 14.5 MT in the years 1941–1967 (Smith 1971). Reports of parasitic-phase lampreys in Lake Erie were sporadic for the first decade following their initial appearance.

The slow rate of establishment in Lake Erie is generally attributed to the limited number of spawning streams with suitable habitat, poor water conditions, and a limited forage base of desirable prey species until more recent times (Lawrie 1970; Pearce et al. 1980). The establishment of sea lamprey in the upper three Great Lakes (Huron, Michigan, Superior), however, was followed by an abrupt and continual decline in commercial lake trout production that was intensified by elevated fishing mortality (Lawrie 1970).

The sea lamprey invasion of Lake Erie and the upper Great Lakes was also followed by extensive changes in the fish communities (Schneider et al. 1996). Prior to sea lamprey invasion, competition, especially in offshore areas, was quite low (Smith 1971). During this time, there were only two apex (i.e., top) predators in the deep waters of the lakes, lake trout and burbot *Lota lota*. All Great Lakes fishes are susceptible to attack by a sea lamprey, but some species are favored over others (Farmer and Beamish 1973). Host habitat, body size, and scale covering influence susceptibility. Sea lampreys tend to occupy cool water and are believed to actively select the largest available host. Thus, large deepwater species, including lake trout, burbot, whitefish, and ciscoes *Coregonus spp.* were the first to decline in the upper Great Lakes (Smith 1972).

The combined effects of sea lamprey parasitism, overexploitation, and other variables such as habitat degradation, led to the ultimate loss of lake trout from lakes Ontario, Erie, Huron, and Michigan, leaving behind only a few populations in Lake Superior, and small remnant populations in Lake Huron (Krueger and Ebener 2004). Although sea lamprey predation was well documented on the aforementioned Great Lakes fishes, it is probable that lake sturgeon were also preyed upon, particularly when abundance of preferred hosts was low. Reports by regional biologists of lake sturgeon with sea lamprey marks and wounds support this hypothesis. The question that remains is whether these attachment events ever lead to mortality in lake sturgeon.

Controlling Sea Lamprey Populations

Because the Great Lakes intersect the political boundary between Canada and the United States, sea lamprey management is a complex endeavor that falls under the jurisdiction of two nations, eight states, one province, and several tribal and First Nation groups. To facilitate coordinated, binational fisheries management, the federal governments of Canada and the United States negotiated and ratified the Convention on Great Lakes Fisheries in 1954 and 1955 respectively. The Great Lakes Fishery Commission was formed in 1956 pursuant to the agreement (www. glfc.org). Since its inception, the goals of the commission have been to control sea lamprey populations in the Great Lakes basin; to coordinate, communicate, and conduct research; and to improve the coordination of fisheries management agencies. The commission works with Fisheries and Oceans Canada, the U.S. Fish and Wildlife Service, the U.S. Army Corps of Engineers, the U.S. Geological Survey, and universities throughout the Great Lakes basin to achieve its sea lamprey control and management goals. The control program operates under an integrated pest management approach and consists of four main components: assessment, barriers, trapping, and lampricides.

The control program has historically relied heavily on chemical methods of control, primarily 3-trifluoromethyl-4-nitrophenol (TFM), costing more than $18 million annually. However, the development of alternative control methods was a keystone of the commission's vision for the past decade and the resulting effort has set the stage for deployment of one or more new alternate control methods in this decade. The sea lamprey control program, in association with stocking, has permitted the successful rebuilding of lake trout populations in Lake Superior and has reduced the impact on other native fishes in all of the Laurentian Great Lakes.

Sea Lamprey and Lake Sturgeon

Little is known about the extent or the effects of sea lamprey predation on lake sturgeon in the wild. The unique morphological traits of lake sturgeon make it probable that the effects of an attack vary greatly compared to effects on common teleost fishes. First, lake sturgeon are armed with rows of hard, bony plates called scutes (a defense mechanism evolved to provide protection from predation in general), which intuitively seem difficult to penetrate. Scutes are much sharper and more plentiful on smaller, juvenile sturgeon and become more rounded and less prevalent as the fish matures. The sharp scutes offer more protection from predation to young sturgeon when their blood volume is less and their chance of dying due to a sea lamprey attack is increased. Sea lamprey are thought to be a size-selective predator, choosing the largest available host on which to feed (Farmer 1980; Cochran 1985; Swink 1990). Increased body size means increased blood volume; however, increased body size does not necessarily allow for better survival following a sea lamprey attack. Collection of lake trout carcasses in Lake Ontario from 1982 to 1985 (Bergstedt and Schneider 1988) confirmed that sea lamprey predominantly kill large lake trout when the trout are abundant, because of the increased frequency of multiple attachments by lamprey for large hosts (Swink 1990).

Scott and Crossman (1973) recounted an observation of a 70 kg lake sturgeon caught in the Bay of Quinte in October 1969 that, when captured, had 15 large sea lampreys attached to its body and bore scars from several previous attacks. The sturgeon was reported to have been sluggish and in poor body condition, probably due to the loss of blood from attacks. Recent accounts of sea lamprey attacks on lake sturgeon are comparatively rare, in part because abundance of other host species, such as lake trout, has rebounded and abundance of sturgeon is currently low. However, field observations do indicate that lake sturgeon still occasionally serve as a host for parasitic sea lampreys. For example, the rate of sea lamprey wounding on subadult and adult lake sturgeon in the St. Marys River, Michigan-Ontario, from 2000 to 2002 was estimated to be 23 percent (T. Sutton, unpublished data). Sea lamprey predation likely continues to be a threat to the future restoration of lake sturgeon. Results from modeling simulations conducted by Sutton et al. (2003) gave reason to believe that sea lamprey predation could affect lake sturgeon rehabilitation more than initially believed. In response to concerns for lake sturgeon in the Sturgeon River, Michigan, the sea lamprey control program established a protocol in 1989 to protect larval lake sturgeon against mortality caused by lampricide application. Under this protocol, the maximum TFM treatment was reduced to 1.3 times the minimum lethal concentration (MLC). Allowable treatment levels were further reduced in 1998 after results of laboratory toxicity experiments showed larval and age-0 lake sturgeon less than 100 mm in length were particularly sensitive to TFM, the primary chemical used for sea lamprey control in the Great Lakes (Johnson, Weisser, and Bills 1999; Boogaard, Bills, and Johnson 2003).

Boogaard, Bills, and Johnson (2003) suggested that the prohibition of treatment for at least 90 days after lake sturgeon spawning would allow sturgeon to reach this length. As a result, allowable treatment levels were not to exceed 1.0 times the MLC for TFM or 1.2 times the MLC for TFM/2% niclosamide and application could not take place until August 1. This protocol is referred to as the "No-Observable-Effect Concentration" (NOEC) protocol or the "sturgeon" protocol. However, modeling results from Sutton et al. (2003) suggested that this protocol may be counterproductive. Less effective lampricide treatments at lower concentrations may result in more predatory lampreys and, consequently, greater levels of parasitism-induced mortality of adult lake sturgeon and other host species.

Throughout the 50-year history of sea lamprey control, only minimal numbers of dead age-0 lake sturgeon have been observed during treatments of streams with lampricides. More specifically, during 1,800 U.S. stream treatments from 1959 to 2000, only 10 dead age-0 lake sturgeon were observed in over 1,600 assessment collections (Adair and Young 2004). Similarly, only 13 dead age-0 sturgeon have been observed during the history of treatments of Canadian streams, with the last observation occurring in 1997 (M. Steeves, Department of Fisheries and Oceans–Sea Lamprey Control Centre, personal communication).

A field study was conducted by T. Pratt, M. Steeves, and L. O'Connor (Department of Fisheries and Oceans–Sea Lamprey Control Centre) in 2008 to assess survival of age-0 lake sturgeon subjected to regular and sturgeon protocol TFM treatments (M. Steeves, DFO Sault Ste. Marie, personal communication). Age-0 lake sturgeon (mean length 78.7 mm; range 57–101 mm; hatchery and native origin) were caged at three sites: the upper Mississagi River, the lower Mississagi River, and the Spanish River (M. Steeves, DFO Sault Ste. Marie, personal communication). The upper portion of the Mississagi River was treated using the sturgeon protocol (1.2 times MLC), whereas the lower portion was treated using a boost to the normal application strategy (1.5 times MLC). Sturgeon were also caged in the Spanish River, which did not undergo treatment during the study period, to control for the effects of stress caused by the caging on survival. Overall, observed lake sturgeon mortality was extremely low (3 percent) and was equally spread among the three treatments. Sturgeon that did die were all notably smaller in size than the rest of the study fish and presumably died from starvation. Results of this study suggest that lake sturgeon mortality related to lampricide treatment is low. These results were surprising given that laboratory toxicity tests and on-site bio-assays have demonstrated that early life stages of lake sturgeon are sensitive to lampricide treatments (Johnson, Weisser, and Bills 1999; M. Boogard, U.S. Geological Survey, unpublished data). It is important to note that lampricide toxicity varies in relation to environmental factors, including stream pH and alkalinity, and may differ at other sites enough to warrant caution (Le Maire 1961; Dawson, Cumming, and Gilderhuis 1975; Marking and Olson 1975). As a result of these preliminary findings, the Great Lakes Fishery Commission has funded two additional years of field research at 12 additional rivers to further investigate the lethality of lampricide treatments on lake sturgeon in the field

(M. Steeves, DFO Sault Ste. Marie, personal communication) and managers should err on the side of caution in the meantime.

A trade-off exists between protecting larval lake sturgeon from lampricide toxicity and protecting juvenile and adult lake sturgeon from sea lamprey parasitism. The Sutton et al. (2003) model suggested that parasitism-induced mortality by sea lampreys could more negatively influence abundance of subadult and adult lake sturgeon, recruitment to age 1, and reproductive potential than lampricide-related mortality on early life stages. If this were true, the survival of subadult and adult lake sturgeon is more crucial than recruitment of early life stages for long-term population persistence. Obviously, this is only a model and models are based on unknowns, so further study is still necessary to investigate the validity of this prediction.

The parasitism-related mortality estimates used in the sublethal model were based on the results of laboratory experiments that used lake trout as the host. As described above, the different life-history traits between lake trout and lake sturgeon likely mean that lethality would differ between these species. To address this uncertainty, a laboratory-based study was conducted by Patrick, Sutton, and Swink (2009). The goal of this study was to estimate the lethality of a single sea lamprey attack on four size-classes of lake sturgeon. A series of 55 experimental trials were conducted whereby one sea lamprey and one lake sturgeon were placed into a tank together after being weighed and measured. Sea lamprey were allowed to feed on lake sturgeon until detachment or mortality of the host or parasite, and each sea lamprey was only permitted one attachment per host. Blood samples were also collected from the lake sturgeon before and after each trial to determine changes in blood chemistry after a lamprey attack. Changes in lake sturgeon growth and blood chemistry were also monitored; the experimental design is described in Patrick, Sutton, and Swink (2009).

Twenty-six percent (n = 14) of the lake sturgeon attacked by a sea lamprey (n = 55) died. The study showed that even a single sea lamprey attack can significantly affect the survival of lake sturgeon, particularly those less than 650 mm in fork length, probably because of their lower host-to-lamprey weight ratio (Patrick, Sutton, and Swink 2009). Changes in lake sturgeon hematology after a sea lamprey attack indicated that direct mortalities resulted from the onset of acute anemia, or the rapid loss of blood. Hosts that lose an amount of blood equivalent to their blood volume (i.e., fish that "bleed out") generally die within two days, whereas hosts that lose 10 percent or less daily can replace those losses and survive (Farmer, Beamish, and Lett 1977). As a result, host survival depends on percentage of blood loss relative to total blood volume (Edsall and Swink 2001). Of fish that do not die directly from blood loss, many suffer from decreases in body condition and may develop fungal infections due to damage to the skin and can die indirectly from an attack.

The 26 percent sturgeon mortality observed in the Patrick, Sutton, and Swink (2009) study was lower than that reported for other species, such as lake trout or burbot, in containment studies with similar study designs (Swink and Hanson 1989; Swink and Fredricks 2000; Bergstedt, Schneider, and O'Gorman 2001; and Swink 2003). For instance, mortality of lake trout, rainbow trout *Oncorhynchus mykiss*, and burbot attacked by a sea lamprey ranged from 40% to 65% (Swink and Hanson 1989; Swink 1990, 1993; Swink and Fredricks 2000). However, the Sutton et al. (2003) simulation model used an estimated range of mortality of 0–22 percent. The results from the laboratory study show that sea lamprey-induced mortality of lake sturgeon may be greater than 22 percent, especially for smaller (< 650 mm FL) lake sturgeon (Patrick, Sutton, and Swink 2009). Consequently, the results of the laboratory study validated the predictions of the model and suggested that sea lamprey parasitism on subadults and adults could significantly affect lake sturgeon population viability.

Still, it needs to be taken into consideration that this was a contained laboratory study and that, as is the case with all laboratory studies, results must be interpreted with caution. In conclusion, although the results of the model predictions suggest that the no-observable effect treatment protocol for the use of lampricides may not be the optimal strategy for achievement of the rehabilitation of self-sustaining lake sturgeon populations, field validations of the model results are required to further inform management. Unfortunately, obtaining mortality estimates of juvenile and adult lake sturgeon in the wild is an exceedingly difficult task.

Gaining Insights about the Precontrol Situation in the Great Lakes from the Predator-Prey Imbalance in Lake Champlain Today

Although reports of sea lamprey attacks on Great Lakes lake sturgeon are infrequent compared to other host species, reports of sea lamprey attachments to Lake Champlain lake sturgeon are much higher. Brian Chipman, a fishery biologist for the Vermont Department of Fish and Game, led an assessment of Lake Champlain lake sturgeon spawning runs for several years and reports that "every one of the over two dozen sturgeon handled had multiple lamprey wounds."

Similar to the situation in Lake Ontario, the lamprey was not an obvious threat to lake sturgeon in Lake Champlain until recently. This observation suggests that either natural control over lamprey numbers or some form of accommodation between predator and prey had been taking place in these lakes until recently (Christie and Kolenosky 1980). If the sea lamprey is indeed native to Lake Champlain, healthy populations of trout, salmon, and sea lamprey coexisted for thousands of years and then suddenly slipped out of balance in the last 30 years. A 2006 study on sea lamprey movement patterns in Lake Champlain showed that sea lamprey are currently feeding on a wide range of host species; wounds have been reported on virtually every fish species large enough to carry a lamprey. High levels of nonsalmonid wounding are indicative of a predator-prey imbalance; much as was seen in the Great Lakes prior to the advent of the sea lamprey control program (Lowe, Marsden, and Bouffard 2006). It is plausible that sea lamprey parasitism affected lake sturgeon in the Great Lakes when predator-prey imbalance was high. By the 1950s, the combination of overfishing and predation by sea lampreys lead to the extermination of lake trout in the lower lakes and depleted stocks in Lake Superior, with balance not being restored until the 1990s.

Effects of Other Aquatic Invasive Species

PREDATION ON LAKE STURGEON EGGS AND LARVAE

In addition to sea lamprey, a number of other aquatic invasive species may affect lake sturgeon rehabilitation. Many invasive fishes can affect recruitment of native fishes by feeding on their eggs. The round goby *Neogobius melanostomus*, a bottom-dwelling, aggressive fish species that invaded from the Ponto-Caspian region, has been observed feeding on lake sturgeon eggs in the lower St. Clair River, Michigan (Jude 2001; Nichols et al. 2003). Lake sturgeon typically spawn in strong currents over cobble and gravel substrate primarily in river systems (Benson et al. 2006). Spawning on this type of substrate usually offers some level of protection from piscine predation; however, laboratory studies indicate that gobies may be well adapted to foraging in substrates with large interstitial spaces (Chotkowski and Marsden 1999). Gobies are believed to preferentially consume eggs in the presence of alternative prey, such as zebra mussels *Dreissena polymorpha*, due to the high caloric content of eggs and the great energetic cost of breaking open the hard zebra mussel shells (Chotkowski and Marsden 1999). The rusty crayfish *Orconectes rusticus* is another invasive species with opportunistic feeding habits that could potentially feed on lake sturgeon eggs and newly hatched young (larvae) (Auer 2004). As most invasions are relatively recent, more studies are required to fully understand the effects of these species on lake sturgeon.

FACTORS AFFECTING JUVENILE FORAGING SUCCESS

The presence of invasive dreissenid mussels in traditional lake sturgeon habitat may also be detrimental to lake sturgeon. The zebra mussel, which first arrived in Lake St. Clair in 1986, now co-occurs with lake sturgeon over much of the latter's distribution (Herbert, Muncaster, and Mackie 1989). Confounding the presence of the zebra mussel, a second nonindigenous species of dreissenid mussel, the quagga mussel *Dreissena rostriformis bugensis*, appeared in 1989 when one mussel was found in Lake Erie; however, it was not recognized as a distinct species until 1991 (Mills et al. 1996). Other species of fish and crayfish tend to prefer dreissenid mussel-covered habitat when given a choice between it and other substrates because of increased macroinvertebrate abundance (macroinvertebrates are attracted to mussels because of the protection provided by the interstitial spaces created), the protection offered by increased habitat structure, or mussels as a potential food source (McCabe et al. 2006).

Juvenile lake sturgeon, on the other hand, avoided zebra mussel-covered habitat 90 percent of the time in laboratory-conducted habitat choice experiments. Juvenile lake sturgeon feed on macroinvertebrates such as mayflies that burrow into the sand and mud on the lake bottom (e.g., Ephemeroptera: *Hexagenia*); juvenile sturgeon cruise over the bottom and use their protrusible mouth to filter invertebrates out of the sediment. Zebra mussels convert potential foraging grounds from areas of soft sediment to a dense, complex habitat that prevents access to burrowing macroinvertebrates. Further, juvenile sturgeon may not recognize zebra mussel colonies as potential foraging grounds (McCabe et al. 2006). Large lake sturgeon (> 600 mm in total length) are; however, capable of using zebra mussels as a food source.

Because zebra mussels supplement the benthic (i.e., bottom dwelling) food web at the expense of the pelagic (i.e., open water) web, a net gain may occur for sturgeon populations, but this increase in food depends on the successful growth of juveniles to a size category that can use zebra mussels or on behavioral shifts to use the macroinvertebrates inhabiting zebra mussel colonies (McCabe et al. 2006). Managers should consider this change in benthic community when developing lake sturgeon rehabilitation strategies.

Evaluating Effects of Aquatic Invasive Species on Lake Sturgeon

The effects of parasitic-phase sea lamprey on Great Lakes fishes are monitored in several ways. One way is by recording sea lamprey wounds on host fishes to determine the extent of parasitism. Sea lamprey wounding data, such as the wound size and the number of each type and stage of wound, are often collected ancillary to other stock assessment information for a variety of fishes in the Great Lakes. Wounding rates, such as the number of observed wounds per 100 fish or the number of observed wounds per unit observation effort, have been shown to be directly proportionate to sea lamprey attack rates (Eshenroder and Koonce 1984).

A dichotomous key for use in the classification of sea lamprey wounds on lake sturgeon was developed as part of the Patrick, Sutton, and Swink 2009 study (Patrick, Sutton, and Swink 2007). This key was developed based on the classification system used for lake trout for over 25 years (King 1980). The key was revised by Ebener, King Jr., and Edsall (2006) to include images of other Great Lakes species, such as lake whitefish *Coregonus clupeaformis*, cisco *Coregonus artedii*, walleye *Sander vitreus*, Chinook salmon *Oncorhynchus tshawytscha*, and white sucker *Catostomus commersonii*. A unique classification system for lake sturgeon was needed because this species differs morphologically from other Great Lakes species, especially in that lake sturgeon lack scales, which makes a difference in healing rates. For example, laboratory observations showed that Type-A wounds (those that create a pit in the skin of the fish) in lake sturgeon often do not penetrate the musculature because of their thick skin and scutes, and result in a more rapid healing rate (Patrick, Sutton, and Swink 2007). The use of the key in the field should provide increased data collection consistency for sea lamprey wounds on lake sturgeon and help to determine changes in rates of predation on lake sturgeon in the wild.

More research is needed on methods to prevent the spread of exotics, and yet preserve migration corridors for native fishes (Auer 2004). The overarching goal of the Great Lakes Fishery Commission's barrier and trapping research theme area is to "transition from using barriers to deny spawning-phase sea lampreys access to spawning habitat, to using barriers to block and selectively trap sea lampreys, and ultimately, to the development and deployment of barriers that are transparent to non-target fishes and of novel, barrier-free traps effective enough for control purposes" (McLaughlin et al. 2007). Much progress has been made to date, with many barriers today having built-in traps to remove sea lampreys and downstream pools to assist the passage of jumping fish.

Maximizing sea lamprey control, while minimizing effects on nontarget native species, is a challenging balance. For example, many Lake Superior stakeholders are anxious to remove or alter the existing hydroelectric dam at the mouth of the Black Sturgeon River in Canada to enhance walleye spawning. Dam removal would have the added benefit of reestablishing connectivity between the Great Lakes and their tributaries for remnant Lake Sturgeon populations. However, this dam limits production to the lower 17 km of the 100 km river in a waterbody that could be an enormous source of sea lamprey to Lake Superior, so management options need to be very carefully considered

Limiting the area of production reduces the amount of stream that needs to be chemically treated. Clearly, dam removal would increase the monetary requirements of the sea lamprey control program, as lampricide treatment frequency and magnitude would need to be increased. It is also likely that dam removal would negatively affect the ecology of Lake Superior because more parasitic phase sea lamprey would enter the population and prey on the fishes. Yet maintaining the dam presents challenges for migrating fishes. The Great Lakes Fishery Commission sponsored a two-day workshop in Turners Falls, Massachusetts, in January 2009, focused on addressing the unique passage issues associated with walleye and lake sturgeon. The potential exists for the Black Sturgeon Dam to become a valuable test case for evaluating fish passage issues.

Conclusions

Mitigating the effects of aquatic invasive species is an important aspect of lake sturgeon rehabilitation. Currently, the effects of sea lamprey predation on lake sturgeon survival are based on laboratory studies and modeling results. Mortality rates in the wild may differ owing to a variety of factors, including increased stress in a laboratory setting (e.g., fish handling) which may lead to an increase in secondary infection, and the presence of preferred alternative hosts (i.e., salmonines and coregonines) in the wild.

In the Great Lakes, wounds on lake sturgeon are rarely recorded in comparison to wounds on other host species. The possibility exists, however, that wounds are not frequently observed because sea lamprey attacks on lake sturgeon result in mortality. If a lake sturgeon were to die from a sea lamprey attack (or multiple sea lamprey attacks), it would be unnoticeable. It is also equally possible that sea lamprey predation on lake sturgeon is not an issue for lake sturgeon, either because it doesn't occur frequently enough in the wild for it to be a problem or, if it does occur, it rarely results in mortality. In any case, laboratory studies should be validated in the field to provide a better understanding of the effects of sea lamprey predation on lake sturgeon. The collection of wounding data using the newly developed dichotomous key for sea lamprey wounds on lake sturgeon will help address these uncertainties. Although sea lamprey may not be feeding extensively on lake sturgeon today because alternative preferred host species are abundant in the Great Lakes, sea lamprey predation may have contributed to mortality in the past when lake trout abundance was low. Sea lamprey parasitism should be considered a potential threat to lake sturgeon survival, especially in the context of modified sea lamprey control treatment protocols that strive to reduce use of lampricides. Further investigation is also needed to determine the effects of other aquatic invasive species on lake sturgeon rehabilitation. The importance of considering the effects of aquatic invasive species in lake sturgeon rehabilitation plans will become increasingly important if the rapid rate of new species introduction into the Great Lakes continues.

REFERENCES

Adair, R. A., and R. J. Young. 2004. Standard operating procedures for application of lampricides in the Great Lakes Fishery Commission integrated management of sea lamprey (*Petromyzon marinus*) control program. USFWS Sea Lamprey Control, Marquette, Michigan, Special report 92-001.4.

Applegate, V. C. 1950. Natural history of the sea lamprey (*Petromyzon marinus*) in Michigan. U.S. Fish and Wildlife Service Special Scientific Report—Fisheries 55.

Aron, W. I., and S. H. Smith. 1971. Ship canals and aquatic ecosystems. Science 174:13–20.

Auer, N. A. 2004. Conservation. *In*: Sturgeons and paddlefish of North America. G. T. O. LeBreton, F. W. H. Beamish, and R. S. McKinley, eds. Kluwer Academic Publishers.

Bailey, R. M., and G. R. Smith. 1981. Origin and geography of the fish fauna of the Laurentian Great Lakes basin. Canadian Journal of Fisheries and Aquatic Sciences 38:1539–1561.

Benson, A. C., T. M. Sutton, R. F. Elliott, and T. G. Meronek. 2006. Biological attributes of age-0 lake sturgeon in the lower Peshtigo River, Wisconsin. Journal of Applied Ichthyology 22:103–108.

Bergstedt, R. A., and C. P. Scheinder. 1988. Assessment of sea lamprey (*Petromyzon marinus*) predation by recovery of dead lake trout (*Salvelinus namaycush*) from Lake Ontario, 1982–85. Canadian Journal of Fisheries and Aquatic Sciences 45:1406–1410.

Bergstedt, R. A., C. P. Schneider, and R. O'Gorman. 2001. Lethality of sea lamprey attacks on lake trout in relation to location on the body surface. Transactions of the American Fisheries Society 130:336–340.

Bigelow, H. B., and W. C. Schroeder. 1948. Cyclostomes. *In*: Fishes of the Western North Atlantic. Sears Foundation for Marine Research Memoir 1.

Boogaard, M. A., T. D. Bills, and D. A. Johnson. 2003. Acute toxicity of TFM and TFM/Niclosamide mixture to selected species of fish, including lake sturgeon (*Acipenser fulvescens*) and mudpuppies (*Necturus maculosus*), in laboratory and field exposures. Journal of Great Lakes Research 29 (Suppl. 1): 529–541.

Bruch, R. M. 1999. Management of lake sturgeon on the Winnebago System—long-term impacts of harvest and regulations on population structure. Journal of Applied Ichthyology 15:142–152.

Bruch, R. M., T. A. Dick, and A. Choudhury. 2001. A field guide for the identification of stages of gonad development in lake sturgeon, *Acipenser fulvescens Rafinesque*, with notes on lake sturgeon reproductive biology and management implications. Publication of Wisconsin Department of Natural Resources. Oshkosh and Sturgeon for Tomorrow.

Bryan, M. B., D. Zalinski, B. Filcek, S. Libants, W. Li, and K. T. Scribner. 2005. Patterns of invasion and colonization of the sea lamprey (*Petromyzon marinus*) in North America as revealed by microsatellite genotypes. Molecular Ecology 14:3757–3773.

Chotkowski, M., and E. Marsden. 1999. Round goby and mottled sculpin predation on lake trout eggs and larvae: Field predictions from laboratory experiments. Journal of Great Lakes Research 25:26–35.

Christie, W. J., and D. P. Kolenosky. 1980. Parasitic phase of the sea lamprey (*Petromyzon marinus*) in Lake Ontario. Canadian Journal of Fisheries and Aquatic Sciences 37:2021–2038.

Cochran, P. A. 1985. Size-selective attack by parasitic lampreys: Consideration of alternate null hypotheses. Oecologia 67:137–141.

Daniels, R. A. 2001. Untested assumptions: The role of canals in the dispersal of sea lamprey, alewife, and other fishes in the eastern United States. Environmental Biology of Fishes 60:309–329.

Dawson, V. K., K. B. Cumming, and P. A. Gilderhuis. 1975. Laboratory efficacy of 3-trifluoromethyl-4 nitrophenol (TFM) as a lampricide. U.S. Fish and Wildlife Service, Investigations in Fish Control 63.

Dymond, J. R. 1922. A provisional list of the fishes of Lake Erie. University of Toronto Studies. Publications of the Ontario Fisheries Research Laboratory 57-73.

Ebener, M. P., E. L. King, Jr., T. A. Edsall. 2006. Application of a dichotomous key to the classification of sea lamprey marks on Great Lakes fish. Great Lakes Fishery Commission Miscellaneous Publlication 2006-02. Http:www.glfc.org/pubs/pub.htm#misc.

Edsall, C. C., and W. D. Swink. 2001. Effects of nonlethal sea lamprey attack on the blood chemistry of lake trout. Journal of Aquatic Animal Health 13:51–55.

Emery, L. 1985. Review of fish species introduced into the Great Lakes, 1819–1874. Great Lakes Fishery Commission Technical Report No. 45. Great Lakes Fishery Commission.

Eshenroder, R. L. 2009. Comment: Mitochondrial DNA analysis indicates sea lampreys are indigenous to Lake Ontario. Transactions of the American Fisheries Society 138:1178–1189.

Eshenroder, R. L., and J. F. Koonce. 1984. Recommendations for standardizing the reporting of sea lamprey wounding data: A report from the Ad Hoc Committee. Great Lakes Fishery Commission. Special Publication 84-1.

Farmer, G. J. 1980. Biology and physiology of feeding in adult lamprey. Canadian Journal of Fisheries and Aquatic Sciences 37:1751–1761.

Farmer, G. J., and F. W. H. Beamish. 1973. Sea lamprey (*Petromyzon marinus*) predation on freshwater teleosts. Journal of Fisheries Research Board of Canada 30:601–605.

Farmer, G. J., F. W. H. Beamish, and P. F. Lett. 1977. Influence of water temperature on the growth rate of the landlocked sea lamprey (*Petromyzon marinus*) and the associated rate of host mortality. Journal of the Fisheries Research Board of Canada 34:1373–1378.

Fetterolf, C. M., Jr. 1980. Why a Great Lakes Fishery Commission and why a Sea Lamprey International Symposium. Canadian Journal of Fisheries and Aquatic Sciences 37:1588–1593.

Finster, J. L. 2007. Investigating injurious species introductions as environmental crimes. M.S. thesis, Department of Fisheries and Wildlife, Michigan State University.

Fortin, R., P. Dumont, and S. Guénette, 1996. Determinants of growth and body condition of lake sturgeon (*Acipenser fulvescens*). Canadian Journal of Fisheries and Aquatic Sciences 53:1150–1156.

Harkness, W. J. K., and J. R. Dymond. 1961. The lake sturgeon: The history of its fishery and problems of conservation. Ontario Department of Lands and Forests, Fish and Wildlife Branch, Maple.

Hay-Chmielewski, E. M., and G. E. Whelan. 1997. State of Michigan lake sturgeon rehabilitation strategy. Michigan Department of Natural Resources, Fisheries Special Report 18.

Herbert, P. D. N., B. W. Muncaster, and G. L. Mackie. 1989. Ecological and genetic studies on *Dreissena polymorpha* (Pallas): A new mollusk in the Great Lakes. Canadian Journal of Fisheries and Aquatic Sciences 46:1587–1591.

Hubbs, C. L., and K. F. Lagler. 1947. Fishes of the Great Lakes region. Cranbrook Institute of

Science Bulletin 26.

Johnson, D. A., J. W. Weisser, and T. D. Bills. 1999. Sensitivity of lake sturgeon (*Acipenser fulvescens*) to the lampricide 3-trifluoromethyl-4-nitrophenol (TFM) in field and laboratory exposures. Great Lakes Fishery Commission Technical Report 62.

Jude, D. J. 2001. Round and tubenose gobies: 10 years with the latest Great Lakes phantom menace. Dreissena! (National Aquatic Species Clearinghouse, SUNY 11(4): 1–9, 12–14.

King, E. L., Jr. 1980. Classification of sea lamprey (*Petromyzon marinus*) attack wounds on Great Lakes lake trout (*Salvelinus namaycush*). Canadian Journal of Fisheries and Aquatic Sciences 37:1989–2006.

Krueger, C. C., and M. Ebener. 2004. Rehabilitation of lake trout in the Great Lakes: Past lessons and future challenges. *In*: Boreal shield watersheds: Lake trout ecosystems in a changing environment. J. Gunn, R. J. Steedman, and R. A. Ryder, eds. CRC Press.

Lawrie, A. H. 1970. The sea lamprey in the Great Lakes. Transactions of the American Fisheries Society 99:766–775.

Le Maire, E. H. 1961. Experiments to determine the effect of pH on the biological activity of two chemicals toxic to ammocoetes. Fisheries Research Board of Canada, Biological Report Series No. 690.

Lowe, D. R., F. W. H. Beamish, and I. C. Potter. 1973. Changes in proximate body composition of the landlocked sea lamprey *Petromyzon marinus* (L.) during larval life and metamorphosis. Journal of Fish Biology 5:673–682.

Lowe, E. A., J. E. Marsden, and W. Bouffard. 2006. Movement of sea lamprey in the Lake Champlain basin. Journal of Great Lakes Research 32:776–787.

Mandrak, N. E., and E. J. Crossman. 1992. Postglacial dispersal of freshwater fishes into Ontario. Canadian Journal of Zoology 70:2247–2259.

Marking, L. L., and L. E. Olson. 1975. Toxicity of the lampricide 3-trifluoromethyl-4-nitrophenol (TFM) to non-target fish in static tests. U.S. Fish and Wildlife Service, Investigations in Fish Control No. 60.

McCabe, D. J., M. A. Beekey, A. Mazloff, and J. E. Marsden. 2006. Negative effect of zebra mussels on foraging and habitat use by lake sturgeon (*Acipenser fulvescens*). Aquatic Conservation: Marine and Freshwater Ecosystems 16:493–500.

McLaughlin, R. L., A. Hallett, T. C. Pratt, L. M. O'Connor, and D. G. McDonald. 2007. Research to guide the use of barriers, traps, and fishways to control sea lamprey. Journal of Great Lakes Research 33 (special issue 2): 7–19.

Mills, E. L., J. H. Leach, J. T. Carlton, and C. L. Secor. 1993. Exotic species in the Great Lakes: A history of biotic crises and anthropogenic introductions. Journal of Great Lakes Research 19:1–54.

Mills, E. L., G. Rosenberg, A. P. Spidle, M. Ludyanskiy, Y. M. Pligin, and B. May. 1996. A review of the biology and ecology of the quagga mussel (*Dreissena bugensis*), a second species of freshwater Dreissenid introduced to North America. American Zoologist 36:271–286.

Nichols, S. J., G. Kennedy, E. Crawford, J. Allen, J. French III, G. Black, M. Blouin, J. Hickey, S. Chernyák, R. Haas, and M. Thomas. 2003. Assessment of Lake Sturgeon (*Acipenser fulvescens*) spawning efforts in the lower St. Clair River, Michigan. Journal of Great Lakes Research 29(3): 383–391.

Patrick, H. K., T. M. Sutton, and W. D. Swink. 2007. Application of a dichotomous key to the classification of sea lamprey marks on lake sturgeon *Acipenser fulvescens*. Great Lakes Fishery Commission Miscellaneous Publlication 2007-02. Http:www.glfc.org/pubs/pub.

htm#misc.

———. 2009. Lethality of sea lamprey parasitism on lake sturgeon. Transactions of the American Fisheries Society 138:1065–1075.

Pearce, W. A., R. A. Bream, S. M. Dustin, and J. J. Tibbles. 1980. Sea lamprey (*Petromyzon marinus*) in the Lower Great Lakes. Canadian Journal of Fisheries and Aquatic Sciences 37:1802–1810.

Ricciardi, A. 2006. Patterns of invasion in the Laurentian Great Lakes in relation to changes in vector activity. Diversity and Distributions 12:425–433.

Rostlund, E. 1952. Freshwater fish and fishing in native North America. University of California Publications in Geography 9:1–314.

Roussow, G. 1957. Some considerations concerning sturgeon spawning periodicity. Journal of the Fishery Research Board of Canada 14:553–572.

Schneider, C. P., R. W. Owens, R. A. Bergstedt, and R. O'Gorman. 1996. Predation by sea lamprey (*Petromyzon marinus*) on lake trout (*Salvelinus namycush*) in southern Lake Ontario, 1982–1992. Canadian Journal of Fisheries and Aquatic Sciences 53:1921–1932.

Scott, W. B., and E. J. Crossman. 1973. The freshwater fishes of Canada. Bulletin of the Fisheries Research Board of Canada 184.

Smith, B. R. 1971. Sea lampreys in the Great Lakes of North America. *In*: The biology of lampreys, vol. 1. M. W. Hardisty and I. C. Potter, eds. Academic Press.

———. 1972. Factors of ecological succession in oligotrophic fish communities of the Laurentian Great Lakes. Journal of the Fisheries Research Board of Canada 29:717–730.

Smith, C. L. 1985. Inland fishes of New York State. Department of Environmental Conservation.

Smith, S. H. 1995. Early changes in the fish community of Lake Ontario, Great Lakes Fishery Commission Technical Report 60.

Smith, B. R., and J. J. Tibbles. 1980. Sea lamprey (*Petromyzon marinus*) in Lakes Huron, Michigan, and Superior: History of invasion and control, 1936–78. Canadian Journal of Fisheries and Aquatic Sciences 37:1780–1801.

Sutton, T. M., and S. H. Bowen. 1994. Significance of organic detritus in the diet of larval lampreys in the Great Lakes Basin. Canadian Journal of Fisheries and Aquatic Sciences 51:2380–2387.

Sutton, T. M., B. L. Johnson, T. D. Bills, and C. S. Kolar. 2003. Effects of mortality sources on population viability of lake sturgeon: A stage-structured model approach. Great Lakes Fishery Commission project completion report.

Swink, W. D. 1990. Effect of lake trout size on survival after a single sea lamprey attack. Transactions of the American Fisheries Society 119:996–1002.

———. 1993. Effect of water temperature on sea lamprey growth and lake trout survival. Transactions of the American Fisheries Society 122:1161–1166.

———. 2003. Host selection and lethality of attacks by sea lampreys (*Petromyzon marinus*) in laboratory studies. Journal of Great Lakes Research 29(Suppl. 1): 307–319.

Swink, W. D., and K. T. Fredricks. 2000. Mortality of burbot from sea lamprey attack and initial analyses of burbot blood. *In*: Burbot biology, ecology, and management. V. L. Paragamian and D. L. Willis, eds. American Fisheries Society.

Swink, W. D., and L. H. Hanson. 1989. Survival of rainbow trout and lake trout after sea lamprey attack. North American Journal of Fisheries Management 9:35–40.

Thresher, R. E. 2008. Autocidal technology for the control of invasive fish. Fisheries 33(3): 114–121.

Trautman, M. B. 1949. The invasion, present status, and life history of the sea lamprey in the waters of the Great Lakes, especially the Ohio waters of Lake Erie. The Ohio State University. The Franz Theodore Stone Laboratory.

Waldman, J. R., C. Grunwald, N. K. Roy, and I. I. Wirgin. 2004. Mitochondrial DNA analysis indicates sea lampreys are indigenous to Lake Ontario. Transactions of the American Fisheries Society 133:950–960.

Welsh, A., T. Hill, H. Quinlan, C. Robinson, and B. May. 2008. Genetic assessment of lake sturgeon population structure in the Laurentian Great Lakes. North American Journal of Fisheries Management 28:572–591.

NANCY AUER

Future Management and Stewardship of Lake Sturgeon

ALDO LEOPOLD IS CONSIDERED TO HAVE BEEN ONE OF THE EARLIEST CONSERVA-
tion activists in the United States and is well known for his best-selling *A Sand County Almanac*, which calls for all humans to live with a "land ethic." Doing so requires that we consider ourselves and all other organisms as valuable and integral partners within an ecosystem community (Leopold 1966). Because he worked for the U.S. Forest Service and wrote of a "land ethic," we often think of Leopold's focus mostly in terms of terrestrial systems. What few know is that some of his first employment obligations concerned aquatic systems and fish (Leopold 1918). His ethic included the aquatic systems within management areas.

His initial position after graduating from Yale was with the U.S. Forest Service in Arizona and New Mexico, where he produced a *Game and Fish Handbook* (1915) calling on foresters to help maintain and support rare wildlife and not just popular game species. He championed the strategy that native fish species should be protected and preferred when considering stocking waters in national forests (Leopold 1918). Leopold's "land ethic" resonated with environmentalists, contributing to the growth of preservation and management programs for parks, reserves, and forests over the last 100 years, yet a similar focus and concern for our aquatic ecosystems was much slower to develop.

All of us can see more and more of the world's natural wild lands being developed for housing, industry, and agriculture as our human population grows, but we are less inclined to realize the extent to which our freshwater resources and the organisms that live in those ecosystems are becoming depleted, polluted, or overexploited, unless there is some drastic outcome of our actions. Examples include the oil and debris spills in the Cuyahoga River in Ohio, which often caught fire in the 1950s and 1960s (Scott 2009) and the large die-offs of the invasive fish—the alewife—in the Great Lakes during the late 1960s (Anon. 1967). The Clean Water Act of 1972 sprang from the deplorable conditions of many lakes and rivers in the 1960s and the need to regulate discharges from industry and municipal sources. The result has been an improvement in water clarity and condition, especially for drinking, but attention seemed to remain focused on water as a resource and not as an integral part of an ecosystem with associated organisms.

As human populations grow around the world, the demand for clean freshwater also grows. Since 2000, much has been written about safety and availability of freshwater resources, and slowly emphasis has been placed on freshwater and marine natural resources. Only within the last 10 to 15 years has the idea of a water ethic been championed (Postel 1997; National Catholic Rural Life Conference 2003), and in March 2008 a conference was held in Santa Clara, California, titled Common Grounds, Common Water: Toward A Water Ethic (proceedings found at *http:// www.internationalwaterlaw.org/bibliography/Ethics/*).

As humans slowly connect the importance of water with that of life, that of both their own *and* plants and animals, we become more aware of what Leopold was suggesting almost 100 years ago—we need to protect all life-forms native to each ecosystem. These organisms have distinct roles within ecosystems that allow them to function and support life best. In the Great Lakes Basin ecosystem, invasive and introduced species such as the zebra and quagga mussels, sea lamprey, and round goby are creating ecological, economic, and industrial havoc, clogging water intakes, and disrupting fish food webs.

Sustainability has become an important term as population growth, demand for resources, climate change, and continued pollution make us all aware of the delicate balance of all of our natural resources. A resource is sustainable if the manner in which we live on earth today allows us to meet our present need of that resource without destroying the ability of future generations to also have sufficient supply and experience of that resource (Garcia and Grainger 1997). In the past, we've often dealt with individual species in small regions but now realize the need to adapt a large ecosystem sustainability strategy.

In the chapters that have preceded, we have learned that the native and unique lake sturgeon has begun to rebound in some regions of its historic range, especially in some portions of the Great Lakes. Few self-sustaining and unrestricted (by dams or habitat loss) groups of lake sturgeon remain, but these groups are rebuilding when conditions allow. As lake sturgeon populations begin to recover through current management practices, all of us need to decide where and how lake sturgeon will be managed and protected. New approaches to sturgeon management can have wider implications for our current approaches to fish management, which include maximum sustainable yields for commercial or sport fisheries or total protection for rare or small fishes. Fisheries management needs to more intentionally include management for sustaining the role of a species within its ecosystem, as Leopold suggested a century ago (Leopold 1918) as well as manage for sustaining the full potential of each species to adapt and evolve, that is, to reduce typical harvest practices of targeting the largest individuals of any population.

Scientists are learning the effects of years of harvesting the largest individuals within a stock—we are actually reducing the overall attainable size of the species (Conover and Munch 2002). Lake sturgeon are the largest fishes in the Great Lakes, and if a fishery continues for those individuals that are the largest throughout the entire ecosystem, we will eventually reduce the potential size of these special fishes. Size matters. The controversy over fishing regulations worldwide that support the catch of the largest individuals must change, or we change the fish itself. Imposing a selective force on any population exposed to fishing will, over time, select for smaller fish, and large trophy fish may become a thing of the past (Conover and Munch 2002; Jorgensen et al. 2007). There are new strategies and scenarios in which better management practices can be applied.

So Why Restore or Protect Lake Sturgeon?

Recently a young woman asked me to explain the benefit of rebuilding lake sturgeon populations. I found it an odd question, as it was couched in the typical thinking that all resources must be seen as a benefit to us. Even 100 years after Aldo Leopold suggested humans need to live harmoniously within our ecosystems and watersheds, underappreciation of the role of humans as part of our environment persists.

There are many stories and descriptions of species now extinct, often as a direct result of human activity. Examples include the passenger pigeon in 1914 (Schorger 1955), Caribbean monk seal in 1952 (Roach 2009), and the Great Lakes deepwater cisco in the 1960s (U.S. Fish and Wildlife Service 2009). Once a species is lost, it cannot be returned. We will never know if any extirpated species held the key to a cancer cure or was a potentially more nutritious food source. But we can be sure it contributed in some way to the ecosystem balance.

What Needs to Be Considered for the Future Management of Lake Sturgeon?

Basic habitat: As discussed in earlier chapters, lake sturgeon need open corridors to reach historic spawning habitat, and those places need good, clean natural freshwater flow to allow eggs to hatch and young to drift downstream to food-rich habitat.

Ecosystem role: Lake sturgeon are benthic feeders; they keep lakes and rivers clean by consuming dead and dying organisms and by feeding on other organisms that help digest and decompose organic debris, thereby helping to maintain ecosystem stability.

Human expectations: Humans have always needed to harvest plants and animals for food. Early Native Americans harvested sturgeon for food, often smoking or drying the flesh. Some Native Americans today still retain sturgeon clan names, and the fish is culturally important to tribes in the Great Lakes region.

Early Europeans who settled North America harvested sturgeon using spears, and in Wisconsin and Michigan small spear fisheries are still allowed. This is because spear fishermen bring economic support to local communities that have a historic record of such activities. Some management agencies also benefit economically by these fisheries, as they supply additional research revenue through the sale of fishing stamps or licenses. In other areas where fishing for sturgeon long ago collapsed, some remnant stocks of fish exist and with some protection may produce a rebound in abundance if habitat is protected and corridors of migration remain open.

As long as there is economic stability, humans may begin to acknowledge and appreciate natural resources not so much for their benefits to us as for their place in the ecosystem. This should create a growing populace satisfied simply to know that an organism as rare and as historic as the lake sturgeon exists and is afforded some protection. As we become educated in the importance of maintaining each organism and allowing at least some members of each organism to attain the size and age they might have without severe human depredations, fewer will want to kill or harvest them with abandon.

New approaches to satisfying man's desire to harmonize with the natural world, often for hunting and fishing, could include encouragement to take photos and releasing fish after capture, which has driven some management agencies to consider catch-and-release options for some popular fishes. On the U.S. West Coast large sport catch-and-release fisheries are growing for white sturgeon. That option is increasing in Michigan waters of Lake St. Clair and Lake Erie as well (M. Thomas, personal communication).

But no matter what approach or strategy managers take for sturgeon, caution is needed. There is evidence that unrestricted catch-and-release programs can produce such fishing pressure that a single sturgeon can be caught many times in one season, and such use of energy (fighting hook capture) may deplete reserves of fat needed to sustain spawning and migration (McKeown 1984). There are also many new, unexpected problems in protecting lake sturgeon populations that may arise: invasive species, poaching, land use changes or sales, dam relicensing or deconstruction, economic imperatives to open lands for mining or other uses, global climate change, and warming waters, bringing increased incidents of disease and parasite spread (Alben et al. 2006).

A Best Practice for Managing This Unique Species?

Historically, fish populations were managed to produce MSY—maximum sustainable yields, again with the aim of retaining sustainable populations. This was done by collecting data on populations thought to have geometric growth (a constant amount of increase) and then harvesting some (usually adults or large individuals) at a rate at which they will be replaced by young fish growing into the adult population, based on life-history data. This type of calculation has not been beneficial to fish stocks because it often overlooked environmental instability (impacts of weather, disease, or predators that can wipe out young fishes), density dependence (when some fish become crowded into one area, their reproductive effort can decrease), and human impacts (lack of complete information on life history, industrial discharges, dams, actual catch). Computer modeling and the ability to handle large data sets have improved some predictions, but uncertainty remains, as ecosystems and organisms are ever changing.

In the last 20 years fishery personnel have been examining additional ways to enhance fish management, especially in the ocean, where many fish stocks have been severely overfished. A relatively new approach is called adaptive management, which has two basic components: to monitor stocks closely and respond quickly. Management plans include constant measuring of important indicators and current status of populations with a planned response, which allows managers to change outcomes (Hilborn and Sibert 1988). Such efforts as closing a fishery when catch quotas are reached or suspending a fishery for a year (or more) if reproductive effort was lost can improve outcomes. Providing opportunities to view fish either in aquariums or at viewing stations can also inform people as to the plight of many fish species and encourage their support of management decisions.

The most recent suggestion to improve management efforts is to move to the ecosystem scale with "nested" governance (Garcia and Grainger 1997). Fishes move between state, country, and oceanic boundaries and their value in trade and economics also changes. Cultural perspectives, local and international law, and general oversight vary, complicating the development of such goals. Walters (2007) suggests large-scale management can work with improvements in three areas. There needs to be an increase in resources for expanded monitoring at such large ecosystem scales, policymakers need to accept uncertainty, and finally, we need fishery managers with vision and a willingness to oversee such projects.

Marine Protected Areas

In October 1972, the U.S. Marine Sanctuaries Program was developed because there was a need to protect spawning grounds and nursery areas of valued marine species. This initiative was later renamed the National Marine Sanctuaries Act, the purpose of which is to conserve, protect, and enhance the biodiversity, ecological integrity, and cultural legacy of marine protected areas.

The official federal definition of an MPA is "any area of the marine environment that has been reserved by federal, state, tribal, territorial, or local laws or regulations to provide lasting protection for part or all of the natural and cultural resources therein" (Executive Order 13158, May 2000). In practice, MPAs are defined areas where natural or cultural resources are given greater protection than the surrounding waters. In the United States, MPAs span a range of habitats including the open ocean, coastal areas, intertidal zones, estuaries, and the Great Lakes. They also vary widely in purpose, legal authorities, agencies, management approaches, level of protection, and restrictions on human uses (*http://mpa.gov/all_about_mpa/basics.html*).

Most MPAs are located within the ocean/marine environment, but there are some in the Great Lakes region that focus on preserving historic shipwrecks. There is a call to develop MPAs for some important Great Lakes fish and aquatic resources. In the U.S. waters of the Great Lakes, as of April 2009, three MPAs had been identified: Isle Royale National Park, Huron Islands National Wildlife Refuge, and Thunder Bay National Marine Sanctuary and underwater preserve (www.mpa. gov). In Canada, one region within the Great Lakes, Georgian Bay, was designated as a UNESCO Biosphere Reserve in 1990 for the purpose of maintaining a balanced relationship between man and the environment. More recently Canada established the Lake Superior National Marine Conservation Area from Thunder Cape to Bottle Point along the Thunder Bay, Ontario, coastline, encompassing one-eighth of Lake Superior to protect and conserve representative examples of Canada's Great Lakes (Parks Canada 2007).

How Are U.S. MPAs Classified?

Because there are many reasons a site may have importance to different "user" groups, there is a five-level system of classification for MPAs and activities that can be conducted within them (Wahle and Uravitch 2006). All have a focus on conservation yet can vary in the level, constancy, permanence, and ecological scale of protection. The focus of any MPA can be centered on natural or cultural heritage conservation (see figure 1, adapted from *http://www.michigan.gov/deq/0,1607,7–135–3313_3677_3701–14531—,00.html*). This may or may not also include a focus on sustainable production—areas established to support extraction of some living renewable resource (fish, plant, etc.). Within these focus areas there can be a variety of levels of protection of resources depending on "use." The protection can range from permanent to temporary, seasonal or year-round, and encompass an entire ecosystem or a single resource within the ecosystem (Wahle and Uravitch 2006).

There is currently some discussion as to the effectiveness of using MPAs as tools to conserve and rehabilitate fisheries (Hilborn et al. 2004; Willis et al. 2003). There is evidence of improvements to fish populations when important stocks are protected within no-harvest areas, and these benefits have been well studied in oceanic environments. Depending on classification and size of the reserve or protected area, some fish and other marine resource populations under study doubled in density, and the average weight of individuals increased three times, with these results seen in as little as three years. Species diversity increased in protected areas by 20 percent, and systems with either several small connected areas or a single large reserve area showed similar improvements (Halpern 2003; Halpern and Warner 2002).

Concern about the success of some reserves can be influenced by the life history of organisms slated for protection as harvestable size adults (large individuals) and pelagic (open water floating) young can regularly move outside the boundaries. Fishing pressure outside reserves can become so intense that improvements in size and abundance seen within the reserve are lost (Hilborn et al. 2004). Walters (2007) believes adaptive fishery management has failed in many protected areas through lack of leadership and oversight of such large projects, lack of willingness to experiment, and lack of funding for careful monitoring programs. Besides needing managers with vision and oversight, new fishery management tactics must involve

Figure 1.
Michigan Marine
Protection Areas

all stakeholders, encompass and model entire ecosystems, and explicitly address the issue of uncertainty (Walters 2007).

A Proposal for Fishery Managers in the Great Lakes with Regard to Lake Sturgeon

We know that lake sturgeon have inhabited and continue to inhabit most regions of the Great Lakes, although in many locations their populations are more limited today than they were historically because of barrier dams and spawning habitat loss. Often, fish management plans have been developed on a lake-by-lake basis or state-by-state basis for species currently deemed sensitive, including sturgeon (Auer 2003; Hoff

2003; Newman, DuBois, and Halpern 2003; Hay-Chmielewski and Whelan 1997). Rarely have Great Lakes fish been considered for complementary management strategies throughout their range or within an entire ecosystem. We propose that such a view and strategy be considered and developed for lake sturgeon within the Great Lakes. Using adaptive strategies, shared by many stakeholders, lake sturgeon management could exemplify new and optimal approaches to ensuring a place for this unique and important species within the entire ecosystem.

Some populations are now so reduced in number that total fishery closures for indeterminate time periods alone may still not result in restoration. Other populations are being slowly and carefully managed to assure rebuilding sustainable stocks, while other populations are utilized for spear or hook-and-line fisheries. Some populations have rebounded so well that the public has begun to catch and release large individuals for the sport of it. Comprehensive management oversight is needed to ensure all stakeholder voices are considered so the lake sturgeon will forever remain at self-sustaining levels throughout the Great Lakes.

MPAs and Managing for Lake Sturgeon

When evaluating potential sites for designation as marine protected areas, managers must consider several criteria. Although evaluations should include social and economic criteria, for this chapter we will limit discussion to several ecological criteria suggested for use in evaluations (Roberts et al. 2003). Roberts et al., gathered as a working group on the Science of Marine Reserves funded by the National Science Foundation, suggest that proposed MPAs should cover many biogeographic regions and habitats, not just a few, thereby providing for all life stages of plants or animals of concern. Sites need to be protected from major human threat, must be of a significant size (or encompass several small sites with corridor connections), and should be focused on sensitive species or species that man exploits. Sites should also retain important ecological services for humans.

With these criteria in mind, we propose that fishery managers consider an MPA as one part of lake sturgeon management in the Great Lakes basin. Lake sturgeon exhibit long-range movements, so effective management will occur only when Canadian and U.S. managers work together. In considering a total ecosystem approach to management and criteria for MPAs, managers must identify other important and sensitive species within the ecosystem that can benefit from designated reserve sites, as suggested in criteria for MPA establishment (Roberts et al. 2003). Using the classification system described earlier (Wahle and Uravitch 2006), we also need to identify within that ecosystem different use approaches such as total preserves, managed historic spear/hook-and-line fisheries, subsistence fisheries and cultural take, and proposed catch-and-release fisheries.

Within the Great Lakes ecosystem, only one nonharvested, self-sustaining population of lake sturgeon remains. That population is found in Lake Superior (Auer 1999). All other populations have, until a very recent closure of the fishery in Ontario in 2008 (COSEWIC 2000; Vélez-Espino and Koops 2008), been subject to harvest—for example, stocks in Lake St. Clair Michigan/Ontario area, and those in the St. Lawrence Seaway. The Lake Superior stock is located in close proximity to one of the few remaining self-sustaining coaster brook trout populations and also an important walleye stock (Hoff 2003). All of these fish are culturally important to local tribes as well as to others for sport and commercial value. They are also key historic members of the Lake Superior ecosystem.

Designing a MPA for this group of fish in south central Lake Superior would include the needed criteria of a broad biogeographic area, incorporation of protection for several sensitive species, and ecosystem services, as suggested by Roberts et al. (2003). The design and classification of subunits within a proposed MPA in Lake Superior for three critical fish species will require collaboration between state (Wisconsin and Michigan), tribal, and possibly Canadian agencies. Within the MPA, sections could be set aside as no-fishing zones or otherwise limit human activity. Areas covered by these protected sections could be different for each species or more inclusive of all species of special concern. These areas would allow at least one stock or population of each species to grow to large size and become monitored sections that contribute to our scientific knowledge on species life history.

An MPA in the Great Lakes region would also provide stock for reintroductions, or simply be a base that "seeds" other regions. It can be managed as a large experiment in freshwater sustainable fisheries, proposed earlier as the future for sustainable fisheries. These strategies are being employed in the oceans and can be adapted to freshwater systems and species. Such a proposal does not limit or restrict establishment of other strategies in other locations within a state or other Great Lake region where current fishing or cultural use may be well established.

Proposal for a Lake Sturgeon / Coaster Brook Trout / Walleye Marine Protected Area

Underwater preserves in the state of Michigan were established in 1980 to provide opportunities for sport diving and to protect historic shipwreck sites (Graf 2009). However, states have also designated underwater preserves to support the conservation of fish and wildlife, such as coral reefs, fishes, and monk seals. Hawaii declared a 100,000-square-mile underwater preserve in June 2006 (Shogren 2006). Such preserves, protecting natural resources, are most often associated with marine systems, usually establishing an area free of heavy harvest pressure. This allows stocks of fish, shellfish, and other organisms to recover and perpetually build populations, which eventually begin to "seed" and restore fisheries in adjacent areas.

Underwater preserves for natural resource protection in the Great Lakes region have not been developed even though human development, exotic species, continued harvest, and barriers to migration corridors have prevented many native aquatic species from rebounding from the overharvest and development in the late 1800s (Auer 1999). Some of the fish species historically abundant and sought either in commercial or sport fisheries, and now absent or low in abundance in the lower Great Lakes, include lean lake trout, ciscoes, coaster brook trout, and lake sturgeon. Lake Superior remains a stronghold for all of these species, and the proposed United States site contains the only self-sustaining, nonharvested population of lake sturgeon in the Great Lakes Basin.

Populations of lake sturgeon in the U.S. waters of the Great Lakes remain either threatened or endangered in all states from continued harvest or reduced habitat availability from dam construction for mills and hydropower (Auer 2004). Although there are areas of free-ranging lake sturgeon (St. Lawrence River, Lake St. Clair, and Detroit River), these populations remain subject to harvest, and spawning sites in river systems have been compromised in most cases. Also, lake sturgeon prefer rivers with a large river mouth delta that provides habitat for production of food organisms. Few unimpacted systems remain today.

Only in Lake Superior and only in the Sturgeon River / Keweenaw Bay system do lake sturgeon move freely to spawn, and then return to the larger lake to feed and rest without threat of harvest or loss of historic habitat. This unique, self-sustaining population should be protected so that one population within the Great Lakes system remains as a case study for other regions and also as a source of material to aid in recovery of the species throughout the rest of the basin. The coaster brook trout, once abundant in Lake Superior, is now restricted to only a few river sites. One is the Salmon-Trout river in Marquette County, Michigan (Newman, DuBois, and Halpern 2003). Coasters moving out of this river need protection from fishing pressure and should be included in a reserve. A population of walleye in the Huron Bay watershed has been selected by managers for rehabilitation (Hoff 2003). The MPA must be of sufficient size to ensure protection for all life stages of these species. Combining agency efforts for an MPA for Lake Superior would be a groundbreaking approach to sustainable fisheries management in freshwater ecosystems.

REFERENCES

Alben, K. T., M. M. Sobiechowska, M. Bridoux, A. Perez-Fuentetaja, and H. Domske. 2006. Climate change and potential for outbreaks of type E botulism: Water temperatures and related physical-chemical conditions associated with past outbreaks. Annual Conference on Great Lakes Research 49.

Anon. 1967. Alewife explosion. Time, July 7.

Auer, N. A. 1999. Lake sturgeon: A unique and imperiled species in the Great Lakes. *In*: Great Lakes Fishery Policy and Management: A Binational Perspective. W. W. Taylor and C. P. Ferreri, eds. Michigan State University Press.

———, ed. 2003. A lake sturgeon rehabilitation plan for Lake Superior. Great Lakes Fishery Commission Miscellaneous Publication 2003-02.

———. 2004. Conservation. *In*: Sturgeons and paddlefish of North America. G. T. O. LeBreton, F. W. H. Beamish, and R. S. McKinley, eds. Kluwer Academic Publishers.

Conover, D. O., and S. B. Munch. 2002. Sustaining fisheries yields over evolutionary time scales. Science 297:94–96.

COSEWIC (Committee on the State of Endangered Wildlife in Canada). 2000. COSEWIC assessment and update status report on the lake sturgeon *Acipenser fulvescens* in Canada. Committee on the State of Endangered Wildlife in Canada.

Garcia, S. M., and R. Grainger. 1997. Fisheries management and sustainability: A new perspective of an old problem? *In*: Developing and sustaining world fisheries resources: The state of science and management, 2nd World Fisheries Congress. D. A. Hancock, D. C. Smith, A. Grant, and J. P. Beumer, eds. Commonwealth Scientific and Industrial Research Organisation.

Graf, T. 2009. Michigan underwater preserves system. Http://www.michigan.gov/deq/0,1607,7-135-3313_3677_3701-14531-,00.html.

Halpern, B. S. 2003. The impact of marine reserves: Do reserves work and does reserve size matter? Ecological Applications 13(1): S117–S137.

Halpern, B. S., and R. R. Warner. 2002. Marine reserves have rapid and lasting effects. Ecology Letters 5:361–366.

Hay-Chmielewski, E. M., and G. E. Whelan. 1997. Lake sturgeon rehabilitation strategy. Michigan Department of Natural Resources, Fisheries Division, Special Report No. 18.

Hilborn, R., and J. Sibert. 1988. Adaptive management of developing fisheries. Marine Policy 12:112–121.

Hilborn R., K. Stokes, J. J. Maguire, T. Smith, L. W. Botsford, M. Mangel, J. Orensanz, A. Parma, J. Rice, J. Bell, K. L. Cochrane, S. Garcia, S. J. Hall, G. P. Kirkwood, K. Sainsbury, G. Stefansson, and C. Walters. 2004. When can marine reserves improve fisheries management? Ocean and Coastal Management 47:197–205.

Hoff, M. H., ed. 2003. A rehabilitation plan for walleye populations and habitats in Lake Superior. Great Lakes Fishery Commission Miscellaneous Publication 2003-01.

Jorgensen, C., and 16 others. Managing evolving fish stocks. Science 318:1247–1248.

Kaufman, L., J. B. C. Jackson, E. Sala, P. Chislom, E. D. Gomez, C. Peterson, R. V. Salm, and G. Llewellyn. 2004. Restoring and maintaining marine ecosystem function. *In*: Defying ocean's end: An agenda for action. L. K. Glover and S. A. Earle, eds. Island Press.

Leopold, A. 1915. Game and fish handbook. U.S. Department of Agriculture, Forest Service, District 3.

———. 1918. Mixing trout in Western waters. Transactions of the American Fisheries Society 47(3): 101–102.

———. 1966. A Sand County almanac. Sierra Club.

McKeown, B. A. 1984. Fish migration. Timber Press.

National Catholic Rural Life Conference. 2003. A water ethic to renew the earth. Www.ncrlc. com/WaterEthic-RenewEarth.html.

Newman, L. E., R. B. DuBois, and T. N. Halpern. 2003. A brook trout rehabilitation plan for Lake Superior. Great Lakes Fishery Commission Miscellaneous Publication 2003-03.

Parks Canada. 2007. Lake Superior National Marine Conservation Area of Canada. www.pc.gc. ca/eng/amnc-nmca/on/super/ne/ne6.aspx.

Postel, S. 1997. Last oasis: Facing water scarcity. W. W. Norton.

Raloff, J. 2001. Underwater refuge. Science News 159:264–266.

Roach, J. 2009. MSNBC science—eight great extinct species. Http://www.msnbc.msn.com/ id/25197251/ns/technology_and_science-science?pg=1#EXTINCTspecies_science.

Roberts, C. M., Andelman, S., Branch, G., Bustamante, R. H., Castilla, J. C., Dugan, J., . . . Warner, R. R. (2003). Ecological criteria for evaluating candidate sites for marine reserves. *Ecological Applications, 13*(1), 199-S214.

Schorger, A. W. 1955. The passenger pigeon. University of Wisconsin Press.

Scott, M. 2009. Cuyahoga River fire 40 years ago ignited an ongoing cleanup campaign. Science news with the Plain Dealer. Http://www.cleveland.com/science/index.ssf/2009/06/ cuyahoga_river_fire_40_years_a.html.

Shogren, E. 2006. Vast Hawaii sea area now a national monument . June 15. Http://www.npr.org/ templates/story/story.php?storyId=5488173.

Vélez-Espino, L. A., and M. A. Koops. 2008. Recovery potential assessment for lake sturgeon (*Acipenser fulvescens*) in Canadian designatable units. Fisheries and Oceans, Burlington, Canada Document 2008/007.

U.S. Fish and Wildlife Service. 2009. Extinct species. Http://www.fws.gov/midwest/Endangered/ lists/extinct.html.

Wahle, C. M., and J. A. Uravitch 2006. A functional classification system for marine protected areas in the United States. www.mpa.gov.

Walters, C. J. 2007. Is adaptive management helping to solve fisheries problems? Ambio 36(4): 304–307.

Willis, T.J., R.B. Millar, R.C. Babcock and N. Tolimieri. 2003. Burdens of evidence and the benefits of marine reserves: putting Decartes before des horse? Environmental Conservation 30(2): 97–103.

About the Authors

BRENDA ARCHAMBO is the founder and president of the Black Lake, Michigan, Chapter of Sturgeon for Tomorrow. Her volunteer leadership established partnerships among the Michigan Department of Natural Resources, stakeholders, and universities to implement research, habitat, conservation, and outreach programs to better understand lake sturgeon reproductive ecology and early life history in order to manage self-sustaining populations of Michigan's lake sturgeon. Archambo is the recipient of the Michigan Public Service Commission Innovative Spirit Volunteer Service Award , the American Red Cross Everyday Heroes Award , the Michigan United Conservation Clubs Special Conservation Award, and the Huron Pines O. B. Eustis Outstanding Individual Conservationist Award. She is also a graduate of the Straits Area Leadership Forum and Great Lakes Fisheries Leadership Institute. Archambo is also an outreach consultant with the National Wildlife Federation and president of the Cheboygan County Economic Development Corporation.

NANCY AUER is an Associate Professor in the Department of Biological Sciences and is also the Department Graduate Program Director at Michigan Technological University. She received her BS, MS, and PhD degrees from the University of

Minnesota Duluth, The University of Michigan, and Michigan Technological University, respectively. Nancy began her journey with lake sturgeon in 1987 using a small nongame wildlife grant from the Michigan DNR to study these fish. She fell in love with them when visiting a laboratory rearing young sturgeon at the University of Wisconsin while working on an identification guide to larval fishes of the Great Lakes. Auer has an active research program in the areas of large lake research and restoration of native fish species. She has published in a variety of well-known journals and has authored several book chapters. She enjoys living in the north country, is an avid birder, and has a home on Lake Superior with her husband Martin and two Welsh Corgies, Peanut and Prairie.

EDWARD A. BAKER is a Michigan native and was an avid angler while growing up in Grand Haven. However, like most people living in the Great Lakes region, he had no idea what a lake sturgeon was. That changed when he entered college and began studying fisheries science. Baker has been a research biologist with the Michigan DNR working in Marquette after finishing graduate school in 1995 and has been studying lake sturgeon since that time. He has studied lake sturgeon distribution and status in Michigan, early life history, population assessment, restoration, streamside rearing, and population demographics. He has authored or co-authored numerous scientific papers on lake sturgeon and serves on a number of interagency committees and working groups. He has also been actively involved in the drafting of various management plans for lake sturgeon in the Great Lakes region.

DAVE DEMPSEY has been an environmental professional since 1982 and is the author or co-author of six previous books, including *On the Brink: The Great Lakes in the 21st Century*, which won a Michigan Notable Book Award. He was environmental advisor to Michigan Governor James Blanchard from 1983 to 1989 and served on the Great Lakes Fishery Commission from 1994 to 2001. He has a master's degree in natural resources policy and law from Michigan State University. He lives in Rosemount, Minnesota.

PIERRE DUMONT first studied biology at the Université de Montréal (1973) and received his MS (1977) and PhD degrees (1996) from Université du Québec à Montréal. He started serving as a fishery biologist in the beginning of the 1970s and was involved in the impact studies of the James Bay hydropower development. He has worked for the Québec government since 1978—mainly in the St. Lawrence River lowlands, the most urbanized part of the province. He is involved in scientific studies on the status and management of lake sturgeon, yellow perch, and American eel, on the long term monitoring of fish communities in the St. Lawrence River, on fish passage and fish habitat improvement, on invasive fish species, and on the recovery of the cooper redhorse, a rare and endangered species endemic to south western Québec. Dumont regularly collaborates with specialists of other national and international organizations of the Great Lakes on water level regulation, fish stocks status and management, and fish passage. He has also been involved in the restoration program of the European sturgeon since 1998, when he had the chance to work at the Cemagref (Bordeaux, France) for one year.

LAURI KAY ELBING attended both Michigan State University and the University of Michigan, studying liberal arts and natural resource policy, respectively, and her work in public policy, political and issue campaigns spans more than twenty years. Perhaps most notable is her work for Congressman John D. Dingell, leading the establishment of the Detroit River International Wildlife Refuge and her work for the National Wildlife Federation developing a high donor advocacy program. She joined the government relations team of The Nature Conservancy in Michigan in March of 2010.

JOHN E. GANNON was born in Detroit, Michigan, in 1942 to a family of automobile factory workers. His career direction was highly influenced by learning to love the out-of-doors while spending summers at his grandparents' cabin on Mullett Lake just south of the Straits of Mackinac in northern Lower Michigan. The lake is part of the inland water route that supports the only inland population of lake sturgeon in Michigan. Gannon has been fascinated with the "big fish" since childhood. He received his BS degree in biology at Wayne State University, MS in fisheries at the University of Michigan, and PhD in zoology (concentration in limnology) at the University of Wisconsin. Gannon worked on the Great Lakes throughout his over four-decades-long career. Following twelve years in academic research and teaching in Michigan and New York, he held various positions in U.S. federal government research and management with the U.S. Geological Survey Great Lakes Science Center (and its predecessor agencies) in Ann Arbor, Michigan, and the International Joint Commission in Windsor, Ontario. Gannon has broad interests in eutrophication, toxic substances, invasive species, habitat protection and restoration, and environmental education and communications. He focused most of his professional career at the interface between research, resource management, and policy. He also enjoyed teaching summers at the University of Michigan Biological Station and Ohio State University's Stone Biological Laboratory. Gannon retired in 2010 and is currently Scientist Emeritus at the USGS Great Lakes Science Center. He also is engaged in various citizen science activities in gardening, birding, and watershed quality.

MARTY HOLTGREN has loved lake sturgeon since he was a youngster when he saw one at the Shedd Aquarium in Chicago. While growing up he lived on the St. Joe River in Niles, Michigan, which once held lake sturgeon but does no longer. He received his bachelor's degree from Bethel College with coursework from Taylor University and the AuSable Institute. He searched to find a graduate position where he could enhance understanding of lake sturgeon to ultimately improve their condition. This led him to Michigan Technological University for a master's degree studying lake sturgeon under the guidance of Dr. Nancy Auer, whom he has worked closely with for the past fifteen years. Currently, he works as the senior fisheries biologist for the Little River Band of Ottawa Indians where he merges science with tribal cultural values, leading to the first streamside rearing facility for lake sturgeon. Holtgren is a PhD candidate at Michigan Technological University. He lives in Manistee, Michigan, with his family (who have spent almost as much time with sturgeon as he has).

ROBERT M. HUGHES was born in 1945 in north central Michigan to working class parents. For the first eighteen years of his life, hunting, fishing, trapping, and a large organic garden were major sources of food for the family. In his early teens, he observed Michigan Department of Natural Resources fish biologists surveying the lake along which he lived and learned that a person could earn a living by studying fish. Hughes obtained multidisciplinary BA and MS degrees at the University of Michigan and a PhD in fisheries at Oregon State University. He conducted research for the U.S. Environmental Protection Agency for thirty-two years as an on-site contractor. He works part-time for Amnis Opes Institute focusing on biological assessments of streams, lakes, and rivers across large geographic extents in the United States, Europe, Brazil, China, and India. Hughes is President-elect of the American Fisheries Society and a member of Oregon's Independent Multidisciplinary Science Team, which reviews state actions for rehabilitating salmon and watersheds.

YVES MAILHOT was born in Sainte-Anne-de-la-Pérade, Québec—a small village along the St. Lawrence River famous for its winter time spawning run and sport fishing festival of Atlantic tomcod. When he was five years old his father took him fishing, which probably contributed to the choice of his future career. He studied biology at the Université du Québec à Trois-Rivières and completed graduate studies in applied ecology at the Université Scientifique et Médicale de Grenoble, in France. Mailhot served as a fishery biologist in the St. Lawrence River for thirty-three years, working mainly for Québec's Ministry of Natural Resources and Wildlife. His lifetime objective was to characterize as precisely as possible what is never seen—the status of the populations of the principal species of fish—in order to contribute to their sustainable exploitation. He developed a unique expertise about the sport and commercial fisheries in the whole St. Lawrence River and collaborated with neighbor colleagues from Ontario and Maritime provinces to national fisheries issues. Mailhot contributed to many scientific committees and to the production of the fishery management plans of the majority of the St. Lawrence River freshwater fish species, like lake sturgeon, yellow perch, American eel, Atlantic tomcod, walleye, sauger, northern pike and striped bass. With his colleague, friend, and co-author Pierre Dumont, Yves has contributed to scientific study and management of the St. Lawrence River lake sturgeon since 1985. They are also particularly proud for creating together a standardized scientific monitoring program of the St. Lawrence River fish communities—the Réseau de suivi ichtyologique (RSI), designed to describe the changes and trends over time of the fish populations and their habitats—in order to help future generations of biologists to manage the resources and evaluate the impact of climate change.

JIMMIE MITCHELL is a Native American treaty rights activist, environmentalist, and Tribal citizen with the Little River Band of Ottawa Indians. Mitchell has actively fought to preserve Tribal hunting, fishing, and gathering rights secured within the 1836 Treaty of Washington. He has been an advocate of Tribal co-management related to existing natural resources and also in the development of programs focusing on the restoration, reclamation, and enhancement of degraded ecosystems and extirpated species significant to the cultural needs of his People. Today, Mitchell is known as a cultural practitioner for his Tribe, serves his Tribe as Natural Resources Director (since 2006), is an active board member on the Chippewa Ottawa Resource Authority (since 2003), and is a representative on the Bureau of Indian Affairs, Midwest Region's Tribal Interior Budget Council.

HOLLY MUIR is a Science Communications Liaison for the U.S. Geological Survey Great Lakes Science Center in Ann Arbor, Michigan. Her Master's project involved assessing host size selection and lethality of sea lamprey on lake sturgeon in a laboratory setting under the advisement of Dr. Trent Sutton at Purdue University. Prior to attending Purdue, she worked with lake sturgeon on Lake of the Woods and Rainy River as an Ontario Ministry of Natural Resources summer STEP student, gillnetting sturgeon for the Minnesota Department of Natural Resources and the Rainy River First Nations Lake Sturgeon Recovery Project in 2004.

HENRY A. REGIER was born in 1930 in the bush of Northern Alberta in a log house built by his immigrant father the previous year from local timber and lumber. A precautionary version of sustainable practice in communitarian living came with that kind of childhood. He was educated at Queen's, Toronto, and Cornell Universities, implicitly in an ecogenic mindscape, as he notes in retrospect. He served as a director of the Institute for Environmental Studies (1989–1994) and Professor in the former Department of Zoology (1966–1995) at the University of Toronto. As an empirical scientist, policy administrator, and practical activist, he has worked collaboratively at all level of governance from local to global with respect to the human population, fisheries, commercialization, chemical pollution, and climate change. The ecosystem approach to the Laurentian Great Lakes Basin that he shares with colleagues integrates all of those issues in ecogenic ways. Among other honors, he is recipient of the Centenary Medal of the Royal Society of Canada, the Conservation Award of the Federation of Ontario Naturalists, the American Fisheries Society Award of Excellence, and the Lifetime Achievement Award from the International Association of Great Lakes Research.

TRENT M. SUTTON is a Professor of Fisheries Biology and the Chair of the Undergraduate Fisheries Program at University of Alaska Fairbanks (UAF), where he has been a faculty member since June 2007. Prior to this he was a faculty member at Lake Superior State University and Purdue University. Since 2001, he has conducted research on lake sturgeon in the Great Lakes, focusing on various aspects of their distribution and movement patterns, habitat use, and life history. Most notably, his research focused on the impacts of various biotic and abiotic stressors (e.g., predation, parasitism, discharge events) on the survival and recruitment of various life stages within the context of long-term population viability.

AMY WELSH has been working on lake sturgeon genetics since 2001. She received a BS from the University of Maryland. While working on a master's of forensic science degree from the George Washington University, she discovered the use of genetics in identifying sturgeon species represented in batches of caviar. She then went on to receive her PhD from the University of California–Davis studying lake sturgeon genetic population structure. She was an Assistant Professor at SUNY–Oswego, continuing genetic studies in the Great Lakes. Currently, Welsh is an Assistant Professor at West Virginia University researching fish and wildlife genetics. She lives in Morgantown, West Virginia, with her husband and two daughters (who have quickly become sturgeon enthusiasts).